INTERNATIONAL CENTRE FOR MECHANICAL SCIENCES

COURSES AND LECTURES - No. 135

PERIODIC OPTIMIZATION

EDITED BY
A. MARZOLLO

VOLUME I

COURSE HELD AT THE DEPARTMENT
OF AUTOMATION AND INFORMATION
JUNE 1972

UDINE 1972

SPRINGER-VERLAG WIEN GMBH

Originally published by Springer–Verlag Wien–New York in 1972

ISBN 978-3-211-81135-1 ISBN 978-3-7091-2652-3 (eBook)

DOI 10.1007/978-3-7091-2652-3

LIST OF CONTRIBUTORS

JAMES E. BAILEY	University of Houston, Houston, Texas, U.S.A.
SERGIO BITTANTI	Politecnico di Milano, Milano, Italy
KUN S. CHANG	University of Waterloo, Waterloo, Ontario, Canada
PETER DORATO	University of Colorado, Colorado Springs, Colorado, U.S.A.
MAGNE FJELD	SINTEF, Trondheim, Norway
GIORGIO FRONZA	Politecnico di Milano, Milano, Italy
MARINO GATTO	Politecnico di Milano, Milano, Italy
GUIDO GUARDABASSI	Politecnico di Milano, Milano, Italy
EUGENE L. LAWLER	University of California, Berkeley, California, U.S.A.
ARTURO LOCATELLI	Politecnico di Milano, Milano, Italy
CLAUDIO MAFFEZZONI	Politecnico di Milano, Milano, Italy
FRANCESCO MAFFIOLI	Politecnico di Milano, Milano, Italy
DINO MANDRIOLI	Politecnico di Milano, Milano, Italy
PIERO MIGLIARESE	Politecnico di Milano, Milano, Italy
SERGIO RINALDI	Politecnico di Milano, Milano, Italy

P R E F A C E

As the reader of this text will understand, the practical problems of industrial and economic nature of even everyday life, which may be modelled as "periodic optimization problems" are so many and so diverse that the existing theories reflect such a variety both in their formulation and in the fields of mathematics from which the methods of solution are taken.

Only a collective and coordinated effort of many researchers may therefore give a rather complete picture of the presently existing approaches to the solution of classes of periodic optimization problems, and a comprehensive theory into which the various problems fit.

This effort has been done by the 15 authors of text, which is the written record of the lectures they gave in the advanced school on "Periodic Optimization" held in the week starting June 5, 1972 at the Department of Autamation and Information of CISM in Udine, Italy.

The promptitude with which the authors have handled their manuscripts, the care with which the technical staff of CISM have typed and printed them make it possible these two volumes to appear within a few months from that week, which all of us remember with pleasure.

I hope that the availability of this text is a stimulus both for applied mathematicians and engineers for further research in such a promising field.

This first volume contains all contributions treating periodic optimization of finite states systems and discrete systems, plus a final paper on necessary and sufficient conditions for the existance and uniqueness of periodic optimal controls for continuous systems, whereas the second volume contains the remaining contributions concerning continuous systems.

The friendship and collaborative atmosphere which exist both among authors and technical staff of CISM is such that a thanks to one representative of each group is to be considered for the whole group. I may therefore mention Sergio Rinaldi for the constance with which he used his competence in coordinating the various contributions, and Enzo Ceschia for his unique enthousiastic industry in taking care of various problems connected with the printing of the volumes.

Angelo Marzollo

Udine, June 1972

GUIDO GUARDABASSI

ARTURO LOCATELLI

SERGIO RINALDI (*)

WHAT IS PERIODIC OPTIMIZATION ? (**)

(*) Istituto di Elettrotecnica ed Elettronica, Politecnico di Milano, Milano, Italy
(**) Supported by Centro di Teoria dei Sistemi (CNR)

1. Introduction

Periodic Optimization is a somewhat new branch of Optimal Control Theory.

Loosely speaking, two main streams of investigations may be recognized within the framework of Optimal Control Theory: Stationary Optimization and Dynamic Optimization.

As well known, Stationary Optimization refers to the problem of selecting the optimal steady state of a given dynamical system and is accounted for a huge number of practical applications, essentially due to the three following reasons. First, the products of industrial processes with characteristics constant in time are usually preferred, secondly, steady state operations are, apparently, the simplest to be implemented and, finally, they can often be selected through a simple optimization procedure.

On the contrary, Dynamic Optimization refers to the problem of determining the optimal transient of a given dynamical system from a preassigned initial state to a suitable final state. Typical applications may hence be found in processes where transitions from a steady-state (eventually optimal) operation to another one occur in an economically relevant way, or even where steady-state operation is a-priori of no interest (for instance, tracking systems).

As a matter of fact, almost all of the control and decision problems arising in the operation of a physical process are, to a great extent, characterized by the requirement of implementing a (possibly optimal) regime. This is why the very idea of Periodic Optimization may be said to be basically originated by the following question: can the optimal steady-state operation be improved by implementing a suitable periodic regime?

Indeed the answer to this question turns out to be affirmative in a number of cases as it can easily be ascertained by the following illustrative examples.

<u>Farmer's problem</u>

By sure, this is one of the oldest decision problems among those which have been faced by farmers in planning cultivations. Suppose the farm is constituted by a certain number of fields where it is possible to grow a certain number of products. How to maximize the productivity of the enterprise? The optimal steady-state policy consists in growing always in the same field the same product. However, it is well known that this is far from being the optimal policy. In fact, since a few thousands of years, farmers do "apply" Periodic Optimization by practising the rotation of crops.

<u>Driver's problem</u>

Suppose your car had a break down and that one of

its wheels is in a hole. You have to get the car out of the hole. Then, the optimal stationary action apparently consists in pushing the car with your strength always in one direction. However, there are circumstances where you will not succeed by simply doing this way. In fact, the most effective action consists in alternatively pushing and pulling the car (periodic action) following its natural oscillations.

Pricing problem

If the optimal solution of the pricing problem were the stationary one, then prices should be constant for long periods of time. However, it is matter of common experience that for some consumers' goods - tipically items sold in large department stores - prices are cyclically lowered (special sales) for short periods of time. By assuming clever decisions in the management of such activities, the conclusion can be drawn that the maximum profit is achieved by periodic pricing.

Dispatching problem

Consider the energy production process for a large area. One of the most important problems is that of selecting what groups of machines must be operated in order to fit with the required amount of energy. Since the diagram of required energy has essentially a (daily) periodic form, it is rather conceivable that a cyclic activation of suitable groups of machines should result in a better over all performance than living always in activity, with different loads, a certain number of groups.

Next section aims to present a rather general mathematical formulation of the problem of Periodic Optimization. To this regard, it has to be noted that Periodic Optimization makes sense only for time-invariant or periodic systems. In the former case, the period of the optimal solution may often be considered free, while in the latter case it has to be chosen, in general, among the multiples of the period of the system even if it is usually a priori fixed to be equal to it.

2. Formal Problem Statement

Since the most significant results in Periodic Optimization are relative to time-invariant systems, the here given mathematical formulation makes reference to the time-invariant case only.

A time-invariant dynamical system Σ

$$\Sigma = \left(T, U, X, Y, \Omega, \Gamma, \Phi(\cdot;\cdot,\cdot), \eta(\cdot,\cdot)\right) \tag{2.1}$$

is given, where T, U, X, Y are, respectively, the sets of admissible times, inputs (controls), states and outputs, Ω and Γ are the sets of admissible input and output functions, and $\Phi(\cdot;\cdot,\cdot)$ is the transition map with the property

$$x(t) = \Phi\left(t;\, x(0), u(\cdot)\right) \qquad \left(t, x(0), u(\cdot)\right) \in T \times X \times \Omega, \tag{2.2}$$

while $\eta(\cdot,\cdot)$ is the output map with the property

$$y(t) = \eta(x(t),\ u(t)) \qquad x(t),\ u(t) \in X \times U \tag{2.3}$$

The optimization problem consists in extremizing the performance index

$$c(\tau, u_{[0,\tau)}(\cdot),\ y_{[0,\tau)}(\cdot)) \triangleq \frac{1}{\tau}\ \mathcal{C}(u_{[0,\tau)}(\cdot),\ y_{[0,\tau)}(\cdot)) \tag{2.4}$$

with respect to the period τ and the control and output functions $u_{[0,\tau)}(\cdot)$ and $y_{[0,\tau)}(\cdot)$, subject to constraints (2.2) and (2.3) and to

$$\Phi(\tau\,;\, x(0),\ u_{[0,\tau)}(\cdot)) = x(0) \in \mathscr{X}_0 \subset X \tag{2.5}$$

$$\tau \in \mathcal{T} \tag{2.6}$$

$$v(\tau, u_{[0,\tau)}(\cdot),\ y_{[0,\tau)}(\cdot)) = \frac{1}{\tau}\ \mathcal{V}(u_{[0,\tau)}(\cdot), y_{[0,\tau)}(\cdot)) = 0 \tag{2.7}$$

Typical features of the problem are both constraint (2.5) and the form of performance index (2.4) and constraint (2.7), which explicitly show the dependence upon the inverse of the time interval τ to be determined.

Such a formulation suggests some general remarks.

(i) Invariance of system (2.1) together with the constraint (2.5) imply the possibility of reproducing in successive intervals of length τ the optimal control $\overset{\circ}{u}_{[0,\tau)}(\cdot)$, supplied by the solution of the problem above. The control which results from this procedure is the optimal periodic solution. If, however, either the invariance of Σ or the constraint (2.5) (substituted, for instance, by an analogous condition on the output) is missing, it is no longer possible, in general, to talk of "periodic optimization".

(ii) No general condition concerning the existence of the solution is available. In fact, when dealing with continuous systems a number of problems seems to point out that, often, a solution does not exist in the classical sense. It has rather to be found in a more general context, through introducing the notion of generalized solutions of differential equations. A different situation occurs when finite systems are concerned. In fact, it generally happens that the optimal value of the performance index for $\mathcal{T}=\{\bar{\tau}\}$ turns to be a decreasing fucntion of $\bar{\tau}$. Thus, if $\mathcal{T}$ is unbounded, this again leads to the conclusion that the optimal solution does not exist.

(iii) The simplest periodic optimization problem is the one with $\mathcal{X}_0 = X$, τ unconstrained and no constraint (2.7). In fact, if this is the case, at least one optimal solution (if any) is constituted by an elementary cycle (i.e. a cycle passing no more than once through the same state).

3. Historical Background

As it can be appreciated through the examples discussed in the introduction, it is difficult to establish a precise date of birth for Periodic Optimization.

Nevertheless, starting from a few decades ago, it is possible to identify in a rather sharp way a small number of independent scientific contributions concerning particular aspects of the problem.

For instance, within the framework of finite state systems, the earliest contributions to Periodic Optimization may be recognized in the first papers dealing with the Travelling Salesman Problem [1] . However it has been necessary to wait until the appearance of the almost contemporary papers by Dantzig et al. [2], Lawler [3] and Romanovskii [4] in order to have treatments of the problems of finding optimal cycles in a graph which were somewhat more germane to the philosophy of Periodic Optimization.

An entirely different kind of contributions may be found in chemical process control starting from the pioneering works by Cannon and his coworkers at the Pennsilvania State University [5]. More precisely, the basic idea raised by these researchers was that of finding proper periodic regimes better than the steady-state one (controlled cycling). More recently, the number of applications in this direction has grown up so rapidly as to justify the 1966 American Chemical Society Symposium on Cyclic Processing Operations. All these application-oriented researches have also been a noteworthy stimulus for theoretical investigations in the field. Thus, as for discrete and continuous dynamical systems, the first steps towards a theory of Periodic Optimization can probably be identified in [6], [7] and [8] respectively.

Finally, it has to be noted that, in spite of the considerable and recent increase of groups of researchers involved in Periodic Optimization, still there exists a relevant inadequacy of the theory in comparison with the number of existing applications. This fact which does not take place in other fields of Optimal Control, constitute, in some sense, a typical aspect of Periodic Optimization. It is just the aim of this series of lectures to point out the status of the art and, contemporarily, stimulate new efforts in the field.

REFERENCES

[1] M. BELLMORE, G. NEMHAUSER: "The Travelling Salesman Problem: A Survey", Opns. Res., vol. 16, pp. 538-558, (1968).

[2] G.B. DANTZIG, W.O. BLATTNER and M.R. RAO "Finding a Cycle in a Graph with Minimum Cost to Time Ratio with Application to a Ship Routing Problem" Theorie des Graphes Int. Symp. Roma, Italy, July 1966, Dunod, Paris.

[3] E.L. LAWLER "Optimal Cycles in Doubly Weighted Directed Linear Graphs" Theorie des Graphes Int. Symp. Roma, Italy, July 1966, Dunod, Paris.

[4] I.V. ROMANOVSKII: "Optimization of Stationary Control of a Discrete Deterministic Process", Kibernetika, vol. 3, No 2, pp. 66-78, March - April 1967 (translated as Cibernetics).

[5] T.T. SZABO, W.A. LLOYD, M.R. CANNON, S.S. SPEAKER: "Controlled - Cycling Extraction", Chemical Engineering Progress, vol. 60, No 1, pp. 66-70, January 1964.

[6] R. ARIS, G.L. NEMHAUSER, D. J. WILDE: "Optimization of Multistage Cyclic and Branching Systems by Serial Procedures" A. I. Ch. E. Journal, vol. 10, No 6, pp. 913-919, November 1964.

[7] L. FAN, C. WANG: "The Discrete Maximum Principle", John Wiley and Sons, New York, 1964.

[8] P.J.M. HORN and R.C. LIN " Periodic Processes: a Variational Approach" I and EC. Proc. Des. and Dev. vol. 6, n. 1, pp. 21-30, Jan. 1967.

PETER DORATO (*)

ASYMPTOTIC OPTIMIZATION IN FINITE DIRECTED GRAPHS

(*) Department of Electrical Engineering and Computer Sciences College of Engineering, University of California, Berkely

1. Introduction

The basic problem considered here is that of finding the "best path" of infinite length in a directed graph with a finite number of vertices. "Best path" is interpreted to mean that path which minimizes a suitable sum of costs, when a cost is assigned to each arc.

It is assumed that the finite vertices are numbered $1, 2, \ldots, N$. Hence the set of vertices are denoted $V = \{1, 2, \ldots N\}$. Arcs in the graph are denoted (i, j), if the arc is directed from vertex i to vertex j. The cost associated with arc (i, j) is represented by real valued weights, c_{ij}. A path is denoted by its corresponding string of arcs, i.e. $\pi = (i_0, i_1)(i_1, i_2) \ldots (i_{k-1}, i_k)$, and the length (*) of the path π by $|\pi|$. An infinite path is one for which $|\pi|$ is infinite. A cycle is a path such that $i_0 = i_k$ and a proper, or simple, cycle is a cycle for which $i_1, \ldots i_k$ are distinct. This is basically the notation in reference [1].

In order to facilitate the generation of infinite paths the following standing assumption is made.

Assumption 1

From every vertex there exists at least one arc

(*) The length is defined as the number of arcs.

directed outward from the vertex.

There are two basic ways in which a path can be generated. One is where an initial vertex and insuing arcs are specified. This generates a single path and is referred to as an <u>open-loop policy</u>. A second method of generating a path is to specify an arc for each vertex. This is referred to as a <u>close-loop policy</u> and generates a whole family of paths one for each new initial vertex. A closed-loop policy may be denoted by a "feedback mapping" $\gamma(i)$ which represents the next vertex given that the present vertex is i. If $\gamma(i)$ is independent of the number of arcs which have been transversed then $\gamma(i)$ is said to be <u>stationary.</u>

The study of infinite paths revolves around the fact that the <u>payoff</u>, total cost, of such a path involves an infinite sum, i.e.,

$$p = c_{i_0 i_1} + c_{i_1 i_2} + \dots = \sum_{n=0}^{\infty} c_{i_n i_{n+1}} \tag{1.1}$$

and this infinite sum may well be divergent. This is obviously the case if all the weights are positive. In the subsequent sec tions we consider this problem in more detail. Several sub-problems develop depending on the structure of the graph and the modifications made in the payoff function. The following assumption is made for all the problems considered here.

<u>Assumption 2</u>

All the weights c_{ij} are non-negative and

the object is to minimize the pay-off p , given by (1.1), or some suitable modification of (1.1).

2. Graphs with Zero Cost Proper Cycles

If a graph has proper cycles, including cycles of length one, i.e., self-loops, with zero cost, and non zero costs for all other arcs then obviously the only infinite sums which yield finite values are those which correspond to paths which deviate from the zero cycles only a finite number of times. Indeed it is fairly clear that for this class of problems the optimal path from a given initial vertex is simply that path which terminates on a vertex in a zero-cost cycle and minimizes the cost up to this terminating condition. The following assumption guarantees the existence of such a solution for every initial vertex.

Assumption 3

From all vertices i and j there is at least one directed path connecting i and j , i.e., the graph is strongly connected.

The optimal path starting at vertex i and minimum value of payoff denoted u_i , may be computed from the Dynamic Programming equation (*)

(*) The minimization over j is over the set permissible vertices from i .

(2.1) $$u_i = \min_j (c_{ij} + u_j)$$

where $u_i = 0$ for all i corresponding to vertices in the zero cost cycles. An extensive discussion of this type of computation may be found in reference [8]. See also reference [2]. The difficult associated problem of finding all zero cost cycles is discussed in references [3] and [4]. This latter problem is simplified if only self-loops have zero costs.

3. Discounted Cost (*)

When all the weights are positive one cannot apply the above procedure since every infinite path yields an infinite value for p. One way to overcome this problem is to introduce a discount factor α, i.e., to consider the discounted pay-off function

(3.1) $$p = \sum_{n=0}^{\infty} \alpha^n c_{i_n i_{n+1}}$$

where $0 \leq \alpha < 1$. This modification may be made palatable for certain problems where such discounting has an economic interpretation.

For the discounted pay-off (3.1) every infinite path yields a finite value for the pay-off since $|c_{ij}| \leq W$ and

(*) Subsequent to the writing of sections 3 and 4 it came to the authors attention that the results presented here may also be found in reference [12].

$$\left| \sum_{n=M}^{\infty} \alpha^n c_{i_n i_{n+1}} \right| \leq \frac{\alpha^M}{1-\alpha} W \tag{3.2}$$

so that the partial sum from M to ∞ converges to zero with $M \to \infty$ proving the convergence of (3.1). The value of p may be computed iteratively as follows. Let $u_i(n)$ denote the minimum value of the sum

$$\sum_{\ell=n}^{M-1} \alpha^\ell c_{i_\ell i_{\ell+1}} \tag{3.3}$$

Then from a Dynamic Programming argument $u_i(n)$ must satisfy the recursive formula

$$u_i(n) = \min_j \left(\alpha^n c_{ij} + u_j(n+1) \right) \tag{3.4}$$

with

$$u_i(M-1) = \min \left(\alpha^{M-1} c_{ij} \right) \tag{3.5}$$

The optimal sum from 0 to $M-1$ is then given by $u_i(0)$. Note that $u_i(0)$ depends on M, i.e., $u_i(0) = u_i(0;M)$. To obtain p:

$$\lim_{M \to \infty} u_i(0;M) = p \tag{3.6}$$

It should be noted that the Dynamic Programming equations (2.1) and (3.4) yield closed-loop policies.

4. Time Average Cost

Another possibility when the sum (1.1) diverges for all paths is to modify the pay-off function to the Cesaro sum,

$$(4.1) \qquad p = \lim_{M \to \infty} \frac{1}{M} \sum_{n=0}^{M-1} c_{i_n i_{n+1}}$$

For problems where the directed graph represents the dynamics of a sequential machine every passage through an arc represents one unit of time and the sum (4.1) is referred to as a <u>time-average</u> pay-off.

This then yields a finite value of pay-off for all divergent sums which are Cesaro summable (Section 12-17 of reference [10]). One important class of paths for which the sum (4.1) exists consists of all paths which end in an arbitrary repetition of a proper cycle. Traiger and Gill (reference [1]) have shown that with assumptions (1.1) and (2.1) a path of this type is guaranteed to exist, hence there is at least one infinite path which has a well defined value of p . However since not all bounded sequences are Cesaro summable it is possible that an infinite path in a given graph may have a divergent limit for the sum (4.1). One such example is given in reference [1] . The following key theorem for this class of problems is proven in reference [1] .

<u>Theorem 1 (Traiger and Gill)</u>

An infinite path which minimizes a time-average payoff is given by arbitrary repetition of an optimal proper cycle (*) preceded by a finite path from the initial vertex to some vertex of the optimal proper cycle.

If assumptions (1.1) and (2.1) are satisfied such an optimal path is guaranteed to exist. Theorem 1 establishes the link between optimal infinite paths (asymptotic or <u>steady state optimization</u>) and optimal proper paths (<u>periodic optimization</u>). Techniques for the determination of optimal proper cycles have been developed by Dantzig et.al. (reference [3]), Lawler (reference [4]), and Dorato (reference [7]). In reference [3] and [4] arbitrary time measures are allowed for each arc, t_{ij} , and optimal proper cycles are proper cycles which minimizes the ratio

$$\frac{\sum c_{i_n i_{n+1}}}{\sum t_{i_n i_{n+1}}} \tag{4.2}$$

over one cycle. Theorem 1 has been generalized to include arbitrary time measures by Rinaldi (reference [5]).

The following theorem illustrates a link between

(*) An optimal proper cycle is a proper cycle with the smallest time-average over one repetition of the cycle.

time-average and discounted pay-offs.

Theorem 2

If the "time-average" limit (4.1) exists then it is equal to the limit of "discounted pay-off"

$$(4.3)\qquad \lim_{M\to\infty} \frac{1-\alpha}{1-\alpha^M} \sum_{n=0}^{M-1} \alpha^n c_{i_n i_{n+1}}$$

as α approaches one from values less than one.

Theorem 2 is a direct consequence of Abel's theorem (Section 13-21 of reference [10]) and the fact that for $0 \leq \alpha < 1$

$$(4.4)\qquad \frac{1-\alpha}{1-\alpha^M} \sum_{n=0}^{M-1} \alpha^n c_{i_n i_{n+1}} = \frac{1}{1+\alpha+\ldots+\alpha^{M-1}} \sum_{n=0}^{M-1} \alpha^n c_{i_n i_{n+1}}$$

while for $\alpha = 1$ the right sum in (4.4) is a time-average sum. However, as pointed out by Rinaldi (reference [5]), optimal solutions to the time-average problem are not in general solutions to discounted problems for any given α , $0 \leq \alpha < 1$. Thus for example the ultimately periodic optimal path of Theorem 1 yields a different pay-off value for each different initial finite portion when $0 \leq \alpha < 1$ and a discounted pay-off function is used.

Finally it should be noted that Theorem 1 establishes the optimality of stationary closed-loop policies for the infinite path problem with time-average pay-offs, since one can always choose a function $\gamma(i)$ which generates the optimal path of Theorem 1 if assumptions 1 and 2 are satisfied.

As pointed out by Traiger and Gill a length L finite sub-path of an infinite path is not necessarily an optimal path over L arcs. However for large enough L the initial segment of an optimal finite path matches that of an optimal infinite path.

5. Optimal Closed-Loop Policies

If stationary closed-loop policies are assumed from the outset, it is possible to characterize directly the optimality of a particular policy and possible iterative procedures for policy improvements. This is the subject matter of reference [7] . The following assumption is required for this theory.

Assumption 4

The graph is such that all closed-loop policies $\gamma(i)$ result in a single proper cycle.

Assumption 4 will be assumed for the remainder of this section. Unfortunately this is not a simple assumption to test, however this assumption considerably simplies the resulting theory. It is an assumption similar to that of a single ergodic chain often made in the theory of Markov chains, which will be discussed in the next section.

We then have the following theorem.

Theorem 3 (reference [7])

If there exist a function v_i , mapping V into the reals, and referred to as the potential such that

$$\lambda^* \leq v_{\gamma(i)} - v_i + c_{i\gamma(i)} \tag{5.1}$$

is satisfied for all i and all possible feedback policies and

$$\lambda^* = v_{\gamma^*(i)} - v_i - c_{i\gamma^*(i)} \tag{5.2}$$

for policy $\gamma^*(i)$, then $\gamma^*(i)$ is an optimal policy, in the sense of yielding the minimum value of time-average pay-off over proper cycles, and λ^* is the minimum value of the time-average pay off over the cycle generated by $\gamma^*(i)$.

In reference [7] it is shown how a potential function can be computed for a given policy so that (5.2) is satisfied. Thus theorem 3 becomes one way of testing the optimality of a given policy and associated proper cycle.

When the condition of the following theorem are satisfied it provides a method of improving a given non-optimal policy.

Theorem 4 (reference [7])

If $v_i^{(1)}$ and $\lambda^{(1)}$ are the potential and time-average cycle pay-off corresponding to policy $\gamma^{(1)}(i)$, and the inequality

$$\lambda^{(1)} > v^{(1)}_{\gamma^{(2)}(i)} - v^{(1)}_i + c_{i\gamma^{(2)}(i)} \tag{5.3}$$

is satisfied for at least one vertex in the cycle generated by $\gamma^{(2)}(i)$, then the cycle corresponding to $\gamma^{(2)}(i)$, yields a smaller value of time-average payoff than $\gamma^{(2)}(i)$.

The major obstacle in applying theorem 4 is that inequality (5.3) must be satisfied for a vertex in the cycle of $\gamma^{(2)}(i)$. If all the vertices are touched (Hamilton path) by all cycles of $\gamma(i)$, this condition is satisfied, however, a graph such that all $\gamma(i)$ yield Hamilton paths is rather unusual. In Markov chains the corresponding assumption, complete ergodicity, is more natural.

6. Markov Chains and Infinite Paths

When uncertainty in transition from vertex to vertex exists one possible model is a random one. In particular one may assume that the transitions are governed by a controlled Markov chain model, with $p_{ij}(k)$ representing the probability of transition from vertex i to vertex j when decision k is made (*). A suitable payoff function is then

$$p_i = \lim_{M \to \infty} E_i \left\{ \frac{1}{M} \sum_{n=0}^{M-1} c_{i_n i_{n+1}} \right\} \qquad (6.1)$$

where E_i denotes the conditional expextion given $i_0 = i$. It can be shown that (6.1) may be written in vector form as

(*) In general k will represent the vertex one would reach if no randomness were present. Since k may depend on the present vertex it is sometimes denoted k_i.

(6.2) $$\underline{p} = \lim_{M \to \infty} \frac{1}{M} \sum_{n=0}^{M-1} P^n \underline{q}$$

where $\underline{p}$ is a column vector with entries p_i, P is an N by N matrix with entries $p_{ij}(k)$ and $\underline{q}$ is a vector with entries

(6.3) $$q_i = \sum_j p_{ij}(k)\, c_{ij}$$

The existence of the limit in (6.2) once again involves a Cesaro sum, but in this case the terms in the sum have a simple algebraic representation, i.e., the nth power of the (transition) matrix P. A discussion of the optimization problem for Markov chains may be found in reference [9] (Chapter 5). We will not dwell on this theory here except to point out its relation to the deterministic problems previously discussed. The Markov chain for a given k is said to be <u>ergodic</u> (reference [11]) if

(6.4) $$P^n = S + T(n)$$

with S a matrix where every row has the same string of elements, say s_1, s_2 ..., s_N, and $T(n)$ is a matrix which goes to the zero matrix as n goes to infinity. The chain is said to be <u>completely ergodic</u> if in addition all of the entries in S are positive. The following assumption is standard in the control of Markov chains.

Assumption 5

The Markov chain characterizing the random transitions from vertex to vertex is completely ergodic for every possible value of the decision variable k .

Assumption 5 guarantees a finite value for the vector payoff (6.2). Indeed the limit in (6.2) then becomes

$$\underline{p} = \begin{pmatrix} \lambda \\ \lambda \\ \vdots \end{pmatrix} \tag{6.5}$$

where

$$\lambda = \sum_i s_i q_i . \tag{6.6}$$

From Dynamic Programming arguments one obtains (reference [11]) the following theorem.

Theorem 5 (Howard, reference [11])

A policy k_i^* which minimizes a payoff of the form (6.1) has the property that for any other policy k_i

$$\lambda^* \leq \sum_j p_{ij}(k_i)\, v_j - v_i + q_i \tag{6.7}$$

where v_i is a function, mapping V into the reals, satisfying

$$\lambda^* = \sum_j p_{ij}(k_i^*)\, v_j - v_i + q_i \tag{6.8}$$

Note that if the Markov chain is allowed to become "deterministic", i.e. $p_{ij}(k_i) = \delta_{ik_i}$ (*), then condition (6.7) becomes identical to condition (5.1).

Unfortunately the completely ergodic assumption is required to obtain an iterative solution of the optimization equation (6.7) and this assumption is violated if $p_{ij}(k) = \delta_{ik}$. Hence a major conclusion is that the Markov theory cannot be taken to its deterministic limit to obtain results for deterministic graph problems, at least not without careful attention to basic assumptions.

7. A Periodic Optimization with "Set" Uncertainty

It is possible to model uncertainty in vertex transition by simply specifying a set of possible "next" vertices rather than a probability measure. Thus for example one may take $\Gamma(i,j)$ as subset of V characterizing the set of possible "next" vertices given the present vertex i and the desired next vertex is j. Without uncertainty one has $\Gamma(i,\gamma(i)) = \gamma(i)$.

A bound on the minimum value of the time-average

(*) δ_{ij} denotes the discrete delta function, i.e., $\delta_{ij}=1$ if $i=j$, and $\delta_{ij}=0$ otherwise.

payoff over a single cycle is given by the following theorem

Theorem 6

If there is a real number λ and a function v_i, mapping V into the reals, such that

$$\lambda \geq \max_{\gamma \in \Gamma} \left(v_{\gamma(i)} - v_i + c_{i\gamma(i)} \right) \tag{7.1}$$

and if each policy $\gamma(i)$ results in a single proper cycle, then for policy an upper bound on all possible values of time-average payoff is given by

Theorem 6 may be proven by simply adding inequality (7.1) over the cycle resulting from $\gamma(i)$. In particular if the resulting cycle is denoted $(i_0, i_1)(i_1, i_2) \cdots (i_{L-1}, i_0)$ one has

$$\lambda \geq \frac{1}{L} \left(c_{i_0 i_1} + \ldots + c_{i_{L-1} i_0} \right) \tag{7.2}$$

for all possible values of $i_1, i_2, \ldots i_{L-1}$ consistent with the set constraint Γ.

REFERENCES

[1] I.L. Traiger and A. Gill, "On an Asymptotic Problem in Finite Directed, Weighted Graphs," Information and Control, Volume 13, N° 6, December 1968, pp. 527-533.

[2] G.B. Dantzig, "All Shortest Routes in a Graph", Teorie des Graphes, International Symposium, Rome, Italy, Dunod, Paris, 1966.

[3] G.B. Dantzig, W.O. Blattner, M.R. Rao, "Finding a Cycle in a Graph with Minimum Cost to Time Ratio with Application to a Ship Routing Problem", ibid.

[4] E.L. Lawler, 'Optimal Cycles in Doubly Weighted Directed Linear Graphs", ibid.

[5] S. Rinaldi, "Asymptotic Properties of the Cost Functions Associated to any Infinite Path in a Finite Graph", Technical Report N° 69-8, Operations Research House, Stanford University, Stanford, California, September 1968.

[6] A. Gill and I.L. Traiger, "Computation of Optimal Paths in Finite Graphs", Computation Methods in Optimization Problems -2, Academic Press, New-York, 1969.

[7] P. Dorato, "Steady-State Optimal Control of Finite-State Machines", Automatica, Volume 7, 1971, pp. 351-358.

[8] R. Bellman, K.L. Cooke and J.A. Lockett, "Algorithms, Graphs, and Computers", Academic Press, New-York 1970.

[9] A. Kaufman and R. Gruon, "Dynamic Programming", Academic Press, New-York, 1967.

[10] T.M. Apostal, "Mathematical Analysis" Addison-Wesley Pub.

Co., Inc., Reading, Massachusset, U.S.A., 1957.

[11] R. Howard, "Dynamic Programming and Markov Process", Technology Press, M.I.T., J. Wiley and Sons, Inc. , New-York, 1960.

[12] J. Shapiro, "Shortest Route Methods for Finite State Space Deterministic Dynamic Programming Problems", SIAM Jour. Appl. Math., vol. 16, n° 6, November 1968, pp. 1232-1254.

EUGENE L. LAWLER (*)

OPTIMAL CYCLES IN GRAPHS AND THE MINIMAL COST–TO–TIME RATIO PROBLEM

(*) Department of Electrical Engineering, University of Colorado formally with the Polytechnic Institute of Brooklyn

Abstract

Let $G=(N,A)$ be a directed graph, and let c_{ij} and t_{ij} be a "cost" and a "transit time assigned to each arc (i, j). We consider three problems. The Negative Cycle Problem: Does G contain a directed cycle for which the sum of the costs is strictly negative? The Minimal-Cost Cycle Problem: Assuming G contains no negative cycles, find a directed cycle for which the sum of the costs in minimum. The Minimal Cost-to-Time Ratio Cycle Problem: Find a directed cycle for which the sum of the costs divided by the sum of the transit times is minimum.

The first two problems can be solved by adaptations of well-known shortest path algorithms which require $O(n^3)$ computational steps, where n is the number of nodes of G. The minimal cost-to-time ratio problem can be solved by a procedure which uses a negative cycle algorithm as a subroutine. This procedure is essentially $O(n^3 \log n)$ in complexity.

Various generalizations of these problems, in the form of side constraints on the cycles, are indicated.

1. Introduction

Let $G=(N,A)$ be a directed graph where N is the set of nodes and A is the set of directed arcs. Let c_{ij} be a (positive or negative) "cost" and t_{ij} be a "transit time" assigned to arc (i,j).

We consider three problems:

Negative Cycle Problem: Does G contain a directed cycle for which the sum of the costs is strictly negative? And if so, find such a cycle.

Minimal-Cost Cycle Problem: Assuming G contains no negative cycles, find a directed cycle for which the sum of the costs is minimum.

Minimal Cost-to-Time Ratio Cycle Problem: Find a directed cycle for which the sum of the costs divided by the sum of the transit times is minimum.

The first two problems can be solved by adapting well-known shortest-path algorithms for the purpose. We review the algorithms of Bellman-Ford and Floyd-Warshall and indicate their application to these problems.

The minimal cost-to-time ratio problem can be solved by a procedure which uses a negative cycle algorithm as a subroutine. We show that the ratio problem can be solved in a number of computational steps that is a relatively low-order polynomial function of the dimension of the problem. (Essentially $O(n^3 \log n)$ for an n-node graph, compared with $O(n^3)$ for a single shortest path computation.)

We conclude by indicating various generalizations of these problems in the form of side constraints on the cycles.

Remark: (1.1) In this paper we deal exclusively with directed networks. Analogous results and computational methods can be obtained for undirected networks. However, the theory and the computational procedures are necessarily much more sophisticated. E.g. shortest path problems in undirected networks are solved by the matching algorithms of Edmonds [5].

2. Bellman–Ford Shortest Path Algorithm

Let c_{ij} be the "cost" of arc (i,j) if there is such an arc, and $+\infty$ if no such arc exists. Let $c_{ij}=0$ if $i=j$.

Suppose we seek to find a least-costly path from node 1 to each of the other nodes, $i = 2,3,\dots,n$. Let

c_i^* = the cost of a least-costly path from node 1 to node i.

If there are no negative cycles, the values c_i^*, $i = 1,2,\dots,n$, must satisfy the following equations:

$$\left.\begin{aligned} c_1^* &= 0, \\ c_i^* &= \min_{k \neq i} \left\{ c_k^* + c_{ki} \right\}. \end{aligned}\right\} \qquad (2.1)$$

The nonlinear equations (2.1) imply implicit functional relationships between the variables c_i^*. Bellman [1] and Ford [8] have proposed solving these equations by a method of "successive approximations":

$$(2.2) \qquad \begin{cases} c_i^{(1)} = c_{1i}, \\ c_i^{(m+1)} = \min_k \left\{ c_k^{(m)} + c_{ki} \right\} \end{cases}$$

Because $c_{ii} = 0$, for all i, it is easy to see that $c_i^{(1)} \geq c_i^{(2)} \geq c_i^{(3)} \geq \ldots \geq c_i^{(m)} \geq c_i^{(m+1)}$.

We can understand the equations (2.2), and demonstrate convergence to c_i^*, by applying the following interpretation:

$c_i^{(m)}$ = the cost of a least-costly path from node 1 to node i, subject to the constraint that the path contains no more than m arcs.

In the absence of negative cycles, there exists an optimal path to any node i with no more than $n-1$ arcs. (If a path contains n or more arcs, at least one node is repeated, and the path contains a cycle.) It follows that $c_i^{(m)} = c_i^*$, for all and all $m \geq n-1$.

The complexity of the computation can be estimated as follows. It is necessary to compute $c_i^{(m)}$ for $i = 2,3,\ldots,n$ and for $m = 2,3,\ldots,n-1$. Each such computation, by (2.2), requires n additions and $n-1$ comparisons. Thus, approximately n^3 additions and n^3 comparisons all required in all.

3. Negative Cycle Computation

We propose to use the Bellman-Ford algorithm to detect the existence of negative cycles. Although the results we state do not strictly require it, it is reasonable to suppose that the graph G is strongly connected. That is, for any pair of nodes i, j there exists a path from i to j and a path from j to i. If G is not strongly connected, it is best to identify its strongly connected components and analyze the components separately.

We leave the proof of the theorem below as an exercise for the reader.

<u>Theorem 3.1</u>

Let G have a path from node 1 to each of the other nodes. Then G contains a negative cycle if and only if, in (2.2), $c_i^{(n)} < c_i^{(n-1)}$, for at least one $i = 1, 2, \ldots, n$.

Theorem 3.1 suggests that to detect the existence of negative cycles, all that is necessary is to carry out the computation of (2.2) for one additional iteration. Thus, the complexity of the computation remains at approximately n^3 additions and n^3 comparisons.

It may be possible to halt the computation earlier by testing for other conditions which are sufficient to indicate the existence or nonexistence of negative cycles.

One condition is that $c_1^{(m)} < 0$, for any m (there is a negative cycle containing node 1). Another such condition is that $c_i^{(m+1)} = c_i^{(m)}$, for all i , and for any m (there are no negative cycles). Still another set of such conditions is given by the following theorem.

<u>Theorem 3.2</u>

G contains a negative cycle if, in (2.2), $c_i^{(m+1)} < c_i^{(m)}$, for some $m = 1, 2, 3, \ldots, n-1$, and for at least $n\ m$ distinct nodes i .

These sufficient conditions may well enable the computation to be ended earlier, but they do not affect the worst-case bound of n^3 additions and n^3 comparisons.

<u>Remarks</u>

(3.1) We have made no attempt to find a "most negative" negative cycle. Any procedure capable of solving that problem could be applied to solve the traveling salesman problem, as well. (Subtract a sufficiently large number from the length of each arc in the traveling salesman problem. A minimum length Hmiltonian tour is then identified with a most-negative cycle.)

(3.2) Any procedure for solving the assignment problem can be used to solve the negative cycle problem. Solve the assignment problem for the matrix of arc costs, with $c_{ii} = 0$. Then an optimal solution to the assignment problem yields a most negative union of disjoint negative cycles, if negative cycles exist. Note:

Assignment algorithms are generally $O(n^3)$ in complexity.

(3.3) We have not described how to construct a negative cycle, when such a cycle is detected. In the tradition of dynamic programming, we simply assert that by appropriate record-keeping (particularly of the choices of k that give rise to minimum values in (2.2)), one can construct such cycles relatively easily as a byproduct of the computation.

4. Improvement in Efficiency of Bellman—Ford Algorithm

A close examination of the computations implied by equations (2.2) reveals that these computations may not be as efficient as they could be. They do not make use of the best information available at each iteration. For example, $c_i^{(m+1)}$ is computed as a function of $c_1^{(m)}, c_2^{(m)}, \ldots, c_n^{(m)}$, even though $c_1^{(m+1)}, c_2^{(m+1)}, \ldots, c_{i-1}^{(m+1)}$ may already have been computed. Making use of these $(m+1)$st order approximations, when available, might accelerate convergence. This is indeed possible, as has been suggested by J.Y. Yen [2].

Suppose we call an arc (i, j) <u>rightward</u> if $i < j$ and <u>leftward</u> if $i > j$. A path is said to contain a <u>change in direction</u> whenever a rightward arc is followed by a leftward arc or vice-versa. Note that because node 1 is the first node of any path, the first arc is rightward and the first change in direction (if any) must be from right to left.

Let us assign a new interpretation to $c_i^{(m)}$. Let

$c_i^{(m)}$ = the cost of a least-costly path from node 1 to node i , subject to the constraint that the path contains no more than m 1 changes in direction.

The appropriate equations for $c_i^{(m)}$ are:

$$
(4.1) \qquad \begin{cases} c_i^{(1)} = c_{1i}, \quad i = 1,2,\dots,n \\ c_i^{(m+1)} = \min\left\{c_i^{(m)}, \ \min\limits_{k>i}\left\{c_k^{(m+1)} + c_{ki}\right\}\right\}, \quad m \text{ odd.} \\ c_i^{(m+1)} = \min\left\{c_i^{(m)}, \ \min\limits_{k<i}\left\{c_k^{(m+1)} + c_{ki}\right\}\right\}, \quad m \text{ even.} \end{cases}
$$

There exists an optimal path to any node i with no more than n arcs and hence no more than $n-1$ changes in direction. It follows that $c_i^{(n)} = c_i^*$.

Each equation is solved by a minimization over about $\frac{n}{2}$ alternatives, on the average, instead of n , as in (2.2). Accordingly, the computational complexity has been reduced by a factor of approximately two: about $\frac{1}{2} n^3$ additions and $\frac{1}{2} n^3$ comparisons are required.

Another, possibly less important, advantage is that storage requirements are also reduced by a factor of approximately two, since not both $c_i^{(m+1)}$ and $c_i^{(m)}$ must be stored. (As soon as $c_i^{(m+1)}$ is computed, it replaces $c_i^{(m)}$.)

Theorems 3.1 and 3.2 apply to Yen's modification (with (4.1) substituted for (2.2) in the statement of the theorems).

Interestingly, Theorem 3.2 is the key to a still further improvement in efficiency. Yen [15] has shown how the length of the computation can be reduced to approximately $\frac{1}{4} n^3$ additions and $\frac{1}{4} n^3$ comparisons by exploiting the fact that at each iteration one additional value $c_i^{(m)}$ becomes equal to c_i^* and therefore affects the calculations no more thereafter.

5. Floyd–Warshall Shortest Path Algorithm[7]

Now suppose we see k to find a least-costly path from node i to node j, for all ordered pairs i, j. Let

c_{ij}^* = the cost of a least-costly path from node i to node j.

We propose to compute c_{ij}^* by a method of successive approximations, as follows:

$$\left.\begin{aligned} c_{ij}^{(1)} &= c_{ij} \\ c_{ij}^{(m+1)} &= \min\left\{ c_{ij}^{(m)},\ c_{im}^{(m)} + c_{mj}^{(m)} \right\} \end{aligned}\right\} \qquad (5.1)$$

We give $c_{ij}^{(m)}$ the following interpretation :

$c_{ij}^{(m)}$ = the cost of a least-costly path from node i to node j, subject to the constraint that the path does not pass through any of the nodes m, $m+1, \ldots, n$ (i, j excepted)

The equations (5.1) call for minimization over two alternatives. It is not clear that the first of these alternatives $\left(c_{ij}^{(m)}\right)$ is that the desired path does not pass through node m and the second alternative $\left(c_{im}^{(m)} + c_{mj}^{(m)}\right)$ is that it does. With this insight, it becomes easy to justify the equations (5.1) and to see that $c_{ij}^{(n+1)} = c_{ij}^{*}$.

We estimate the complexity of the Floyd-Warshall algorithm as follows. It is necessary to compute $c_{ij}^{(m)}$ for $i = 1,2,\ldots,n$ for $j = 1,2,\ldots,n$ and for $m = 2,3,\ldots,n+1$. Each such computation, by (5.1), requires one addition and one comparison. Thus, exactly n^3 additions and n^3 comparisons are required overall.

6. Min–Cost cycle problem

The theorem below requires no proof.

Theorem 6.1

G contains a negative cycle if and only if, in (5.1), $c_{ii}^{(m)} < 0$, for any $i = 1,2,\ldots,n$ and any $m = 2,3,\ldots,n+1$.

The Floyd-Warshall algorithm has essentially the same upper bound on the number of computational steps as the (unmodified) Bellman-Ford algorithm. However, in practice it appears that the Bellman-Ford algorithm is much more likely to terminate early, and should be preferred for the negative cycle problem.

The Floyd-Warshall method is most useful, for our purposes, because it can be used to solve the minimal-cost cycle problem.

Theorem 6.2

If G does not contain a negative cycle, then the cost of a least-costly cycle is given by

$$\min_i \left\{ c_{ii}^{(n+1)} \right\}$$

where $c_{ii}^{(n+1)}$ is determined by (5.1), with $c_{ii} = +\infty$, for all i.

7. Minimal cost–to–time ratio Problem

Recall that in addition to a cost c_{ij} , each arc (i , j) of G is assigned a transit time t_{ij} . We now wish to find a directed cycle C for which the cost-to-time ratio,

$$\frac{\sum_C c_{ij}}{\sum_C t_{ij}} \tag{7.1}$$

is minimal.

As an example of this type of problem, consider the following situation suggested by Dantzig [3] .

A tramp steamer is free to choose its ports of call and the order in which it calls on them. A voyage from port i to port j earns p_{ij} dollars profit, and requires t_{ij} days time (including time for loading cargo at i and unloading at j). What ports should the steamer visit, and in what order, so as to maximize its mean daily profit?

The system can be represented by a labelled directed graph, with nodes identified with ports (states) and arcs with possible voyages (state transitions). Each arc has associated with it a profit (output) $p_{ij} = - c_{ij}$ and a transit time t_{ij} .

A solution to the optimization problem is found by identifying a directed cycle reachable from the initial node for which the ratio of total profit to total travel time is maximized. The tramp steamer then sails from its starting point to any port in this cycle, and then continues to sail around this cycle indefinitely.

In order to avoid complications, and to avoid having to examine a multiplicity of cases, we assume that the sum of transit times around any cycle is strictly positive. I.e., for all C ,

$$\sum_C t_{ij} > 0 \qquad (7.2)$$

Let us guess a minimum value λ for the ratio (7.1). Suppose we give each arc (i, j) in G a new cost $\bar{c}_{ij} = c_{ij} - \lambda t_{ij}$. There are three situations that may exist with respect to these modified cost coefficients $\bar{c}_{ij}$:

<u>Case 1.</u> There is a negative cycle in G.

<u>Case 2.</u> There is no negative cycle, and the cost of a minimal-cost cycle is exactly zero.

<u>Case 3.</u> There is no negative cycle, and the cost of a minimal-cost cycle is strictly positive.

Suppose that Case 1 holds. Let C be a negative cycle. Then

$$\sum_C \bar{c}_{ij} = \sum_C (c_{ij} - \lambda t_{ij}) < 0 .$$

By assumption (7.2), it follows that

$$\frac{\sum_C c_{ij}}{\sum_C t_{ij}} < \lambda .$$

In other words, the guessed value of λ is too large and C is a cycle for which the cost-to-time ratio is strictly smaller than λ.

By similar analysis, we see that in Case 2 the guessed value of λ is exactly equal to the minimal cost-to-time ratio, and in Case 3 λ is too large.

These observations suggest a search procedure based on the testing of successive trial values of λ . There are two principal types of searches we might conduct: "monotonic" and "binary".

One can organize a monotonic search by choosing an initial value $\lambda^{(0)}$ at least as large as the minimum cost-time ratio, and then, by successive applications of Case 1, obtain $\lambda^{(0)} > \lambda^{(1)} > \lambda^{(2)} > \ldots$, until some $\lambda^{(p)}$ is found for which Case 2 Holds. This must occur in a finite number of steps, because G contains only a finite number of cycles. Hence there are only a finite number of possible cost-time ratios.

In fact, it can be shown by the analysis in Section 8 that a monotonic search requires only $O(n^3)$ trial values of λ . Since each trial value requires an $O(n^3)$ negative cycle computation, the monotonic search procedure is $O(n^6)$ overall.

The binary search [10] provides a much better bound and is probably much more effective in practice. The binary search proceeds as follows. Suppose we know that the optimum cost-time ratio is contained in the interval (a,b). We first try the trial value $\lambda = \frac{a+b}{2}$. If Case 1 holds, we know that the optimum ratio is contained in the interval $\left(a, \frac{a+b}{2}\right)$. If Case 3 holds, the optimum ratio is in the interval $\left(\frac{a+b}{2}, b\right)$. If Case 2 holds, we have, of course, found the optimum value of the cost-time ratio at the first try.

We continue in this way, halving the remaining interval with each trial value. After k trial values the length of the remaining interval can be no greater than $(b-a)/2^k$. Or, to put it another way, the number of trial values required to reduce the length of the interval to ε is $\log_2(b-a) - \log_2 \varepsilon$.

We continue the interval-halving procedure until the remaining interval is so small that only one distinct cost-time ratio can be contained within the interval. Then one additional trial value of λ is sufficient to find the cycle with this minimal ratio. (The final value of λ is chosen to be equal to the largest value in the ε-interval; either Case 1 or Case 2 must hold.)

In the next section we indicate how to determine the values a , b , and ε , estimate the length of the computation.

Remark.

(7.3) It is not difficult to formulate the ratio cycle problem as a " fractional" linear programming problem. I.e. a problem with linear constraints and an objective function that is the ratio of two linear forms. However, the standard techniques for solving such problems, e.g. [2] , are unworkable, because they do not yield integer solutions.

8. Complexity of Cost–to–time Ratio Algorithm

We suppose that all parameters c_{ij} and t_{ij} are

integers and that $|c_{ij}| \leq \gamma$ and $|t_{ij}| \leq \tau$, for all i, j. Each cycle contains at most n arcs, and the smallest possible value for the sum of the transit times around any cycle is unity. Hence the minimum and maximum attainable cost-to-time ratios are $a = -n\gamma$ and $b = n\gamma$, respectively.

Furthermore, if the cost-to-time ratios for two cycles are unequal, those ratios must differ by at least $\varepsilon = 1/n^2\tau^2$.

It follows that the binary search procedure outlined in Section 7 requires no more trial values of λ than

$$\log_2(b-a) - \log_2\varepsilon = \log_2(2n\gamma) - \log_2(1/n^2\tau^2) = 1 + 3\log_2 n + \log_2\gamma\ 2\log_2\tau.$$

Thus, the number of negative-cycle problems which must be solved is $O(\log n + \log\gamma + \log\tau)$.

Suppose we are concerned with graphs of various sizes, but with similar cost and time parameter values, if we assume that γ and τ are invariant with n (or even if γ and τ increase as polynomial functions of n) then the number of negative cycle computations is simply proportional to $\log_2 n$ and the overall computations is $O(n^3\log_2 n)$.

Even if we make no assumptions about the nature of the parameters c_{ij} and t_{ij}, it is clear that the number of computational steps is bounded by a polynomial function in the number of bits required to specify an instance of the problem. In this sense, the minimal cost-to-time ratio problem can be considered to be a well-solved combinatorial optimization problem.

Remarks

(8.1) The cost-to-time ratios for two cycles, if unequal, must differ by at least $1/n^2\tau^2$. However, it is possible that the ratios for two closed paths (not cycles) may differ by less than this amount. Thus, the final interval of length ε may contain a multiplicity of realizable ratios, with respect to closed paths other than cycles. However, the minimum ratio is realized by a cycle, and the negative-cycle computation enable us to find such a cycle.

(8.2) The present author [10] and Shapiro [12] had previously proposed an $O(n^4)$ algorithm for solving the ratio problem, in the special case that $t_{ij} = 1$, for all i, j. This was based on the recursion

$$c_{ij}^{(1)} = c_{ij} \quad \text{where} \quad c_{ii} = +\infty$$

$$c_{ij}^{(m+1)} = \min_k \left\{ c_{ik}^{(m)} + c_{kj} \right\}$$

One then computes

$$\min_{i,m} \left\{ \frac{c_{ii}^{(m)}}{m} \right\}$$

to obtain the minimum cost-to-time ratio. However, except in very special circumstances it appears that this method is inferior to the binary search method.

9. Side Constraints

It is interesting to consider variations of the cost-to-time ratio problem. For example, suppose there are only m ports, $m < n$, at which the tramp steamer may refuel. A closed path must be found which is optimal, subject to the constraint that the travel time between two successive refuelings does not exceed a certain time T. This version of the ratio problem can be solved as follows.

We first observe that, because of the refueling restrictions, an optimal closed path may not be a cycle. That is, it can pass through a given refueling port no more than once, but it may pass through a given nonrefueling port as many as times, but no more than once between successive refueling ports. It follows that the minimum and maximum attainable cost-to-time ratios are $a = -m(n-m+1)\gamma$ and $b = m(n-m+1)\gamma$, and that the cost-to-time ratios of two closed paths can be assumed to differ by at least $\varepsilon = 1/m^2(n-m+1)^2\tau^2$. A small bit of algebra shows that, although the number of trial values of λ is greater than in the unconstrained problem, this number is $O(\log n + \log\tau + \log\gamma)$ as before.

For each trial value of λ, we must test for the existence of a negative closed path, subject to refueling constraints. This can be done in two stages. First $c_{ij}(T)$ is computed, for all refueking ports i and j, where

$c_{ij}(t)$ = the cost of a least-costly path from node i to node j, subject to the constraint that the path requires no more than t units of time.

Then a conventional negative cycle computation is carried out for the m-node network of refueling ports, with arc lengths $c_{ij}(T)$. We assert that there is a negative closed path in the original n-node network, subject to refueling constraints, if and only if there is a negative cycle in the m-node network.

One way to carry out the computation of $c_{ij}(T)$ is as follows.

Assume that each t_{ij} is a positive integer. Then we have

$$\left.\begin{aligned} c_{ij}(t) &= +\infty && \text{if } t < 0 \\ &= \min_k \left\{ c_{ik}(t - t_{kj}) + c_{kj} \right\} && \text{if } t \geq 0, \end{aligned}\right\} \quad (9.1)$$

where we assume a self loop with $c_{ii} = 0$, $t_{ii} = 1$, at each node i. Equations (9.1) can be viewed as a generalization of equations (2.2), and require $O(mn^2T)$ computational steps. (Incidentally, note that c_{ij} is actually $\bar{c}_{ij} = c_{ij} - \lambda t_{ij}$, where λ is the current trial value.)

An alternative computation is as follows.

$$
(9.2)\qquad
\begin{cases}
c_{ij}(t) = +\infty & \text{if } t < 0 \\
= \min\limits_{k \neq i,j} \min\limits_{0 \leq \delta \leq \tau} \left\{ c_{ik}\left(\left\lfloor \frac{t}{2} \right\rfloor - \delta\right) + c_{kj}\left(\left\lceil \frac{t}{2} \right\rceil + \delta\right) \right\}, & \text{if } 0 \leq t < t_{ij} \\
= \min \; c_{ij}, \min\limits_{k \neq i,j} \min\limits_{0 \leq \delta \leq \tau} \left\{ c_{ik}\left(\left\lfloor \frac{t}{2} \right\rfloor - \delta\right) + c_{kj}\left(\left\lceil \frac{t}{2} \right\rceil + \delta\right) \right\}, & \text{if } t \geq t_{ij}
\end{cases}
$$

(The symbols "$\lfloor\ \rfloor$" and "$\lceil\ \rceil$" denote "greatest integer less than or equal to" and 'least integer greater than or equal to," respectively.) An analysis of equations (9.2) shows that they imply $O(m\,n^2 \tau^2 \log T)$ computational steps to obtain $c_{ij}(T)$, for all m refueling ports i and j .

If we take the number of trial values of λ to be essentially $O(\log n)$, then the overall computation is either $O\ mn^2 T \ \log\ n$) or $O(mn^2\tau^2 \log T \log n)$, depending upon whether equations (9.1) or (9.2) are used. If either T or τ is sufficiently small, the computation may not be too forbidding.

<u>Remark:</u>

(9.3) The author is indebted to S. Rinaldi for suggesting improvements in the steamer refueling computation, and for pointing out that the monotone search method for the ratio problem is $O(n^6)$ in complexity.

REFERENCES

[1] R.E. Bellman, "On a routing Problem", Quart. Appl. Math., XVI (1958) 87-90

[2] A. Charnes and W.W. Cooper, "Programming with Linear Fractional Functions", Naval Research Logistics Quarterly, 9 (1962) 181-186.

[3] G.B. Dantzig, W. Blattner and M.R. Rao, "Finfing a Cycle in a Graph with Minimum Cost to Time Ratio with Applications to a Ship Routing Problem", Theory of Graphs, International Symposium, Dunod, Paris, and Gordon and Breach, New York (1966) 77-83.

[4] S.E. Dreyfus, "An Appraisal of Some Shortest Path Algorithms", Operations Research, 17 (May 1969) 395-412.

[5] J. Edmonds, "Path, Trees, and Flowers", Can. J. Math., 17 (1965) 449-467.

[6] M. Florian and P. Robert, "A Direct Search Method to Locate Negative Cycles in a Graph", Management Science, 17 (1971) 307-310.

[7] R.W. Floyd, "Algorithm 97, Shortest Path", Comm. of the ACM, 5 (1962) 345.

[8] L.R. Ford, Jr, "Network Flow Theory", the RAND corp., P-923 (Aug. 1965).

[9] B.L. Fox, "Finding Minimal Cost-Time Ratio Circuits", Operations Research, 17 (1969) 546-551.

[10] E.L. Lawler, "Optimal Cycles in Doubly Weighted Directed Linear Graphs", Theory of Graphs, International Symposium, Dunod, Paris, and Gordon and Breach, New York (1966) 209-213.

[11] E.F. Moore, "The Shortest Path Through a Maze", Proc. Int. Symp. on the Theory of Switching, Part II, April 1967, The Annals of the Computation Laboratory of Harvard University, 30, Harvard University Press (1959).

[12] F. Shapiro, "Shortest Route Methods for Finite State Space Dynamic Programming Problems", SIAM J. Applied Math, 16 (1968) 1232-1250.

[13] I.L. Traiger and A. Gill, "On an Asymptotic Optimization Problem in Finite, Directed, Weighted Graphs," Information and Control, 13 (1968) 527-533.

[14] J.Y. Yen, "An Algorithm for Finding Shortest Routes from All Source Nodes to Given Destination in General Networks", Quart. of Appl. Math., 27, (Jan. 1970) 526-530.

[15] J.Y. Yen, "A Shortest Path Algorithm", Ph.D. Dissertation, University of California, Berkeley, California (1970).

ARTURO LOCATELLI

PIERO MIGLIARESE

SERGIO RINALDI (*)

CONSTRAINED PERIODIC OPTIMIZATION OF FINITE AUTOMATA (**)

(*) Istituto di Elettrotecnica ed Elettronica, Politecnico di Milano, Milano, Italy

(**) Supported by Centro di Teoria dei Sistemi (CNR)

Abstract

The problem of Periodic Optimization of finite automata is here considered with reference to the case where either "path constraints" or "integral constraints" are present. Algorithms are presented for both problems and structural properties of their solutions are discussed. The theory here developped is shown to fit with significant applications.

1. Introduction

Given a finite automaton (U, X, δ) where U and X are the input and state sets and δ is the next-state function $X \times U \to X$, i.e.,

$$x(t+1) = \delta\,(x(t), u(t)) \tag{1.1}$$

the problem of Periodic Optimization can be formulated as follows:

$$\min \frac{1}{\tau} \sum_{t=0}^{\tau-1} c\,(x(t), u(t)) \triangleq \min \frac{1}{\tau}\,\mathcal{C} \triangleq \min C \triangleq C^{\circ} \tag{1.2}$$

subject to (1.1) and the constraints

$$x(0) = x(\tau) \in \mathcal{X}_0 = \text{a subset of } X \tag{1.3}$$

(1.4) $\tau \in \mathcal{T}$ = a subset of the integers

(1.5) $$\frac{1}{\tau} \sum_{t=0}^{\tau-1} v^{(k)}(x(t), u(t)) = \bar{V}^{(k)}, \quad k = 1, \ldots, q.$$

As it is well known, such a problem can be described by means of a finite and oriented graph: every node i represents a state and every arc (i, j) a possible transition from state i to state j. Moreover, a cost $c_{ij} \gtrless 0$ and q constraint values $v_{ij}^{(k)}$ are associated to every arc (i, j) and the problem consists in the determination of the cycle with minimum cost per unit time. More precisely, letting S be the set of all cycles Γ in the graph, $\tau(\Gamma)$ and $\mathcal{C}(\Gamma) = \sum_{\Gamma} c_{ij}$ be, respectively, the number of arcs and the cost of Γ, the loss rate (cost to time ratio)

$$c(\Gamma) = \frac{\mathcal{C}(\Gamma)}{\tau(\Gamma)}$$

is associated to every element Γ of S. Thus, the general problem becomes

$$\min_{\Gamma \in S} c(\Gamma) = c^{0}$$

subject to constraints (1.3) - (1.5).

Particular, but interesting, cases of problem (1.1) - (1.5) are those in which only some of the constraints are present. The two problems dealt with in this work refer to the case where constraints (1.3), (1.4) (called path constraints) or constraints (1.5) (called integral constraints) must be verified. A third important constrained problem in Periodic Optimization of finite systems is that of the optimal Hamiltonian circuit. Such a problem can not, however, be viewed as a particular case of the general formulation (1.1) - (1.5) and it is not discussed here since already widely considered in the literature [1] - [4].

It must be noted that, in general, the transition from state i to state j can be accomplished by means of different control actions, thus more than one cost is associated to every pair of states of the automaton. This implies that the corresponding graph is, in general, a multigraph. Nevertheless, only the case of linear graphs is dealt with in this work since, as it is shown later, it is always possible to reduce a given multigraph to a linear graph.

All proposed algorithms for the solution of a Periodic Optimization problem, [5] - [8], make reference to the unconstrained case. This is not casual, since, under this assumption the set S can be reduced to the finite subset S^* of the elementary cycles.

The existing algorithms apply to the more general case where the transition $i \rightarrow j$ have not the same lastings. Therefore, also a time $t_{ij} > 0$ is associated to each arc (i,j) If the lasting of a cycle is defined as $\tau(\Gamma) \triangleq \sum_{\Gamma} t_{ij}$, then an optimal cycle (if any) is still constitued by an elementary cycle. Moreover, allowing the t_{ij}'s , not to be equal to one, the multigraph case can be reduced to the simpler linear case. In fact, if r arcs link node i to node j , then it suffices to substitute every arc, but one, from i to j with a couple of arcs passing through an artificial node, with costs, times and constraint values which have half the value of the original ones.

As for the constrained problems, the here proposed algorithms need the t_{ij}'s to be integers. This is not a severe restriction, since the problem with rational lastings can be easily reformulated in terms of integer lastings. Furthermore, an analogous simplifying assumption requiring that all costs c_{ij} and constraint values $v_{ij}^{(k)}$ be rational, is made in the following.

2. Path Constraints

The problem of Periodic Optimization of finite automata is discussed in this section relative to the particular case of path constraint.

More precisely, the problems is as follows:

(2.1) $$\min_{\Gamma \in S} \frac{\mathcal{C}(\Gamma)}{\tau(\Gamma)} \triangleq \min_{\Gamma \in S} C(\Gamma) \triangleq C^{\circ}$$

under a constraint on the period of the cycle

$$\tau(\Gamma) \in \mathcal{T}$$

and the "path constraint" requiring that the optimal cycle meets a given state which now onwards is called state $1 \left(\mathcal{X}_0 = \{1\}\right)$.

Apart from particular cases, an optimal solution of this problem does not exists if $\mathcal{T}$ is unbounded (see the following remark).

Remark 1

Let $\{\Gamma^0\}$ be the set of all optimal solutions of the problem without path constraint $(\mathcal{X}_0 = X)$ and $\mathcal{T} = \{1,2,\ldots\}$. If state 1 does not belong to any of the cycles Γ^0 but is strictly connected with at least one of these cycles, then the problem with path constraint $(\mathcal{X}_0 = \{1\})$ and $\mathcal{T} = \{1,2,\ldots\}$ has no solution. ▨

Proof

Let $S_1 = \{\Gamma^{(1)}\}$ be the set of cycles passing through state 1. Since state 1 is strictly connected with a cycle $\bar{\Gamma}^0 \in \{\Gamma^0\}$, there exists a cycle $\bar{\Gamma}^{(1)} \in S_1$ with at least one state in common with $\bar{\Gamma}^0$. Then consider the sequence $\{\Gamma_k^{(1)}\}$ where each cycle $\Gamma_k^{(1)}$ is constitued by $\bar{\Gamma}^{(1)}$ and by $\bar{\Gamma}^0$ repeated k times. Thus,

$$c\left(\Gamma_k^{(1)}\right) = \frac{\tau\left(\bar{\Gamma}^{(1)}\right) c\left(\bar{\Gamma}^{(1)}\right) + k\tau\left(\bar{\Gamma}^0\right) c\left(\bar{\Gamma}^0\right)}{\tau\left(\bar{\Gamma}^{(1)}\right) + k\tau\left(\bar{\Gamma}^0\right)}$$

is the average cost of $\Gamma_k^{(1)}$. Since

$$\lim_{k\to\infty} C(\Gamma_k^{(1)}) = C(\bar{\Gamma}^0)$$

and $C(\Gamma_k^{(1)}) > C(\bar{\Gamma}^0)$ for all k , the property is proved. ▨

In wiew of Remark 1 , the problem with path constraint makes sense, in general, only if $\mathcal{T}$ is bounded. Hence, from now on,

$$\mathcal{T} = \{1, 2, \ldots, N\} \tag{2.2}$$

As in the unconstrained case, a simple property can be stated also for the problem with path constraint (the proof is trivial and hence omitted).

<u>Remark 2</u>

At least one optimal solution of the problem with $\mathcal{T}$ as in (2.2) consists of a cycle passing only once through state 1 . ▨

This property will be used in one of the steps of the algorithm presented below, which is mainly based on the following remark (see [5] - [8] for the unconstrained Periodic Optimization Problem).

<u>Remark 3</u>

Consider

$$d_{ij} \triangleq c_{ij} - \varrho t_{ij}$$

as cost of the transition $i \to j$ in the given finite automaton. Let

$$\varrho > C^o$$

where C^o is given by eq. (2.1) and S^- be the set of all cycles Γ (from now on called negative cycles) passing through state 1 , with $\tau(\Gamma) \in \mathcal{T}$ and such that

$$\sum_{\Gamma} d_{ij} < 0 . \qquad (2.3)$$

Then

i) S^- is not empty

ii) $C(\Gamma) < \varrho , \quad \forall \Gamma \in S^-$. ▨

Proof

i) Consider the cycle Γ_1^o which satisfies eq. (2.1) and all constraints. Then

$$\frac{\sum_{\Gamma_1^o} c_{ij}}{\sum_{\Gamma_1^o} t_{ij}} = C^o < \varrho$$

from which

$$\sum_{\Gamma_1^o} d_{ij} < 0 .$$

ii) Eq. (2.3) gives

$$\sum_{\Gamma} c_{ij} - \varrho \sum_{\Gamma} t_{ij} < 0 \qquad \forall \Gamma \in S^- .$$

Hence

$$\frac{\sum_{\Gamma} c_{ij}}{\sum_{\Gamma} t_{ij}} = C(\Gamma) < \varrho \qquad \forall \Gamma \in S^{-}.$$

This remark suggest an iterative procedure for the determination of the optimal solution of problem (2.1). In fact, if $\varrho^{(k)} \geq C^{o}$ is the value of the constant ϱ at the k-th iteration, a value $\varrho^{(k+1)} \leq \varrho^{(k)}$ can be found by looking for a cycle $\Gamma^{(k)}$ with $\sum_{\Gamma(k)} (c_{ij} - \varrho^{(k)} t_{ij}) < 0$. Moreover, $\varrho^{(k+1)} = \varrho^{(k)} = C^{o}$ if such a cycle does not exists; otherwise, $C^{o} \leq \varrho^{(k+1)} < \varrho^{(k)}$. Since the set of feasible cycles is finite, this procedure ends to the optimal cycle in a finite number of iterations. Of course, the major task of each iteration is to find a negative cycle Γ in a graph with costs d_{ij} , passing through node 1 and with $\tau(\Gamma) \in \mathcal{T}$.

<u>Outline of the algorithm</u>

1. Choose $\varrho^{(1)} > C^{o}$ (for example , $\varrho^{(1)} = \max \frac{c_{ij}}{t_{ij}}$).
2. Let $k = 1$.
3. Let $\varrho = \varrho^{(k)}$ and evaluate $d_{ij} = c_{ij} - \varrho t_{ij}$.
4. Find a cycle $\Gamma^{(k)}$ (if any) passing through node 1 and such that

$$\sum_{\Gamma(k)} t_{ij} \leq N \qquad \sum_{\Gamma(k)} d_{ij} < 0 \ .$$

5. If the cycle $\Gamma^{(k)}$ of step 4 does not exist, the cycle $\Gamma^{(k-1)}$ is an optimal cycle and $C^{o} = \varrho^{(k)}$. Otherwise let

$$\varrho^{(k+1)} = \frac{\sum_{\Gamma^{(k)}} c_{ij}}{\sum_{\Gamma^{(k)}} t_{ij}}$$

and go to step 3 with k replaced by $k+1$. ▨

Step 4 of the algorithm may be accomplished by means of Dynamic Programming as follows.

First verify if $d_{11} < 0$ and $t_{11} < N$. If so, the self loop in node 1 is a feasible negative cycle. Otherwise, let $f_i(t)$, $i \neq 1$, be the minimum cost of a path from node i to node 1, of total duration t and not passing through node 1. Then,

$$f_i(t) = \min_{j \neq 1} \left[d_{ij} + f_j(t - t_{ij}) \right], \quad t \neq t_{i1}, \quad i = 2,3,\ldots,n$$

$$f_i(t_{i1}) = \min \left\{ d_{i1}\, ; \min_{j \neq 1} \left[d_{ij} + f_j(t_{i1} - t_{ij}) \right] \right\}, \quad i = 2,3,\ldots,n$$

where

$$f_i(t) = \infty \qquad \forall t \leq 0 .$$

Obviously, a feasible negative cycle passing only once through node 1 (see Remark 2) exists if and only if

$$f_i(t) + d_{1i} < 0 \; , \qquad t + t_{1i} \leq N \tag{2.4}$$

for some i and t. Moreover, if eq. (2.4) is satisfied, the corresponding cycle can be easily determined provided that the computation of the function $f_i(t)$ is suitably organized.

Example

Consider the finite automaton with $n = 5$ described in Fig. 2.1 where the entries correspond to possible state transitions, and suppose $N = 23$.

c_{ij}	1	2	3	4	5
1		5		1	
2	1			2	
3		1		1	
4			1		5
5	5	5			

t_{ij}	1	2	3	4	5
1		1		5	
2	1			1	
3		1		5	
4			5		1
5	1	1			

Fig. 2.1. : The costs C_{ij} and times t_{ij}.

Then, by applying the algorithm, the optimal solution is obtained in six iterations as shown in Fig. 2.2.

Iteration k	$\varrho^{(k)}$	Cycle $\Gamma^{(k)}$	$\sum_{\Gamma^{(k)}} d_{ij}$	$\sum_{\Gamma^{(k)}} t_{ij}$
1	5	(1-2-1)	-4	2
2	3	(1-4-5-1)	-10	7
3	11/7	(1-4-5-2-1)	-4/7	8
4	3/2	(1-4-3-2-1)	-14	12
5	1/3	(1-4-3-4-3-2-1)	-4/3	22
6	3/11	(1-4-3-4-3-2-1)	0	22

Fig. 2.2.: The six iterations necessary to find the optimal solution

From Fig. 2.2 it results that the optimal cycle (1 - 4 - 3 - 4 - 3 - 2 - 1), though passing only once through node 1 , is not elementary. ▨

3. Integral Constraints

The problem of Periodic Optimization of finite automata is discussed in this section with reference to the particular case of integral constraints.

More precisely, the problem is as follows:

$$\min_{\Gamma \in S} \frac{\mathcal{C}(\Gamma)}{\tau(\Gamma)} \triangleq \min_{\Gamma \in S} C(\Gamma) \triangleq C^{0} \tag{3.1}$$

under the equality constraints (the case of inequality constraints is completely analogous)

$$\frac{\mathcal{V}^{(k)}(\Gamma)}{\tau(\Gamma)} \triangleq V^{(k)}(\Gamma) = \bar{V}^{(k)} \qquad k = 1,\dots,q. \tag{3.2}$$

The existence of the solution of problem (3.1), (3.2) is now discussed from a geometrical point of view. Let $\Gamma_1, \dots, \Gamma_k$ be the elementary cycles. The loss rate $C_i \triangleq C(\Gamma_i)$ and the q "constraint rates" $V_i^{(k)} \triangleq V^{(k)}(\Gamma_i)$, $k = 1,\dots,q$ can be associated to each cycle Γ_i, $i = 1, \dots, K$. Thus, K points $Z_1, \dots, Z_K$ representing all loss and constraint rates associated to the elementary cycles can be located in the $(q+1)$-dimensional vector space $V^{(1)} \times \dots \times V^{(q)} \times C$ (see Fig. 3.1).

Since each cycle Γ can be decomposed, in an obvious way, into elementary cycles Γ_i repeated n_i (≥ 0) times, then the loss rate $C(\Gamma)$ and the constraint rates $V^{(k)}(\Gamma)$ are a suitable convex combination of the C_i's and $V_i^{(k)}$'s and

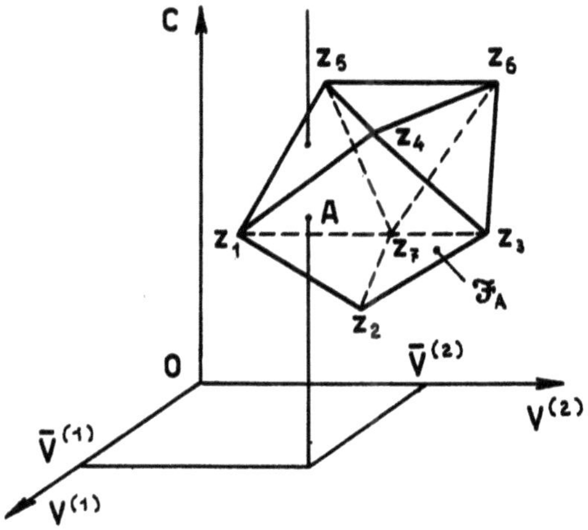

Fig. 3.1.: The Polyhedron P.

the coefficients λ_i, $i=1,\dots,K$ of the convex combination are

$$\lambda_i = \frac{n_i\,\tau(\Gamma_i)}{\sum_{i=1}^{K} n_i\,\tau(\Gamma_i)} . \tag{3.3}$$

Therefore, the polyedron P defined by the convex cover of the K points $z_1,\dots,z_K$ (see Fig. 3.1) contains the set of all loss and constraint rates which can be obtained with non-elementary cycles. The coefficients λ_i defined by eq. (3.3) can not, however, be arbitrary real numbers. Therefore, a cycle can not, in general, be associated to each point of P even if it can be stated that in each neighbourhood of such a point there are points corresponding to non-elementary cycles. Thus, the existence of the optimal solution of the problem is guaranteed if there exists a cycle to be associated to the lower intersection A between the straight line $V^{(k)}=\bar{V}^{(k)}$, $k=1,\dots,q$ and the polyedron P (see Fig 3.1).

Of course, if a cycle Γ is associated to the point A, then it can be decomposed into elementary cycles corresponding to some of the vertices of the face $\mathfrak{F}_A$ of P where the point A lies (see points z_1, z_2, z_3, z_4 of Fig. 3.1).

Moreover, since all data (c_{ij}, t_{ij}, $v_{ij}^{(k)}$, $\bar{V}^{(k)}$) are rational numbers, the point A has rational coordinates which can be expressed as a convex combination - with rational coefficients λ_i - of the coordinates of the vertices of $\mathfrak{F}_A$. This means that the point A is always associated with some of the elementary cycles Γ_i corresponding to the vertices of $\mathfrak{F}_A$ repeated $n_i > 0$ times. Therefore, the solution of the problem does not exist only if these cycles Γ_i do not constitute a cycle Γ.

Nevertheless, the determination of these cycles Γ_i and of the integers n_i is of interest even if the problem does not have a solution. In fact, if C^o is the solution of problem (3.1), (3.2) with the operator "min" substituted by the operator "inf", then it is possible to approximate at will the value C^o of the performance and $\bar{V}^{(k)}$ of the constraints by means of a cycle constituted by some of the elementary cycles Γ_i corresponding to the vertices of $\mathfrak{F}_A$ and by an elementary cycle which simply connects them. Obviously, the cycles Γ_i must be run a sufficiently high number of times in order to obtain a satisfactory approximation of C^o and $\bar{V}^{(k)}$;

in this way, the "connecting" cycle has a negligeable influence on the loss and constraints rates.

Moreover, since the polyedron P belongs to a $(q+1)$-dimensional vector space, the point A can be expressed as the convex combination of at most $q+1$ vertices of $\mathfrak{F}_A$. This implies that C^0 and $\bar{V}^{(k)}$ can always be approximated at will by means of a non-elementary cycle constituted by $q+1$ elementary cycles Γ_i and (if this is the case) by a "connecting" cycle .

The cycles Γ_i corresponding to the vertices of $\mathfrak{F}_A$ and the number n_i of times they are run, can be found by first solving the integer programming problem

$$(3.4) \qquad \min \frac{\sum_{i,j} c_{ij} x_{ij}}{\sum_{i,j} t_{ij} x_{ij}}$$

subject to

$$(3.5) \qquad \sum_i x_{ij} = \sum_k x_{jk} , \qquad j = 1,\dots,n$$

$$(3.6) \qquad \frac{\sum_{i,j} v_{ij}^{(k)} x_{ij}}{\sum_{i,j} t_{ij} x_{ij}} = \bar{V}^{(k)}, \qquad k = 1,\dots,q$$

$$(3.7) \qquad x_{ij} = \text{non negative integer} \qquad i,j = 1,\dots, n$$

and then applying the following procedure.

<u>Procedure 1</u>

Let $x_{ij}^{(1)} = x_{ij}$ and

$$n_1 = \min_{i,j} \left(x_{ij}^{(1)} \mid x_{ij}^{(1)} > 0 \right) \tag{3.8}$$

Moreover, let (i_1 , i_2) be the first arc in the lexicographic ordering such that

$$x_{i_1 i_2}^{(1)} = n_1 .$$

Now, determine the arc (i_2 , i_3) so that

$$x_{i_2 i_3}^{(1)} = \min_{j} \left(x_{i_2 j}^{(1)} \mid x_{i_2 j}^{(1)} > 0 \right) \tag{3.9}$$

and continue this way until a cycle Γ_1 is obtained. Thus, by letting

$$x_{ij}^{(2)} = \begin{cases} x_{ij}^{(1)} & (i,j) \notin \Gamma_1 \\ x_{ij}^{(1)} - n_1 & (i,j) \in \Gamma_1 \end{cases}$$

one can reapply eqs. (3.8), (3.9) and obtain a new cycle Γ_2 and its associated integer n_2.

By iteration, a finite sequence of cycles $\Gamma_1,\ldots,\Gamma_r$ and the associated integers $n_1, \ldots, n_r$ are obtained. ▨

Consider the following Linear Program

$$\min \sum_{i,j} c_{ij}\, y_{ij} \tag{3.10}$$

subject to

(3.11) $$\sum_i y_{ij} = \sum_k y_{jk} \qquad j=1,\dots,n$$

(3.12) $$\sum_{i,j} v_{ij}^{(k)} y_{ij} = \bar{V}^{(k)} \qquad k=1,\dots,q$$

(3.13) $$\sum_{i,j} t_{ij} y_{ij} = 1$$

(3.14) $$y_{ij} \geq 0 \ , \quad i,j=1,\dots,n \ .$$

Since all data are rational numbers, an optimal solution y_{ij}^{o} of (3.10) - (3.14) can be given the form

(3.15) $$y_{ij}^{o} = \frac{x_{ij}^{o}}{T}$$

where T is a positive integer and all x_{ij}^{o} are non-negative integers. Thus, eq. (3.13) gives

(3.16) $$\sum_{i,j} t_{ij} x_{ij}^{o} = T \ .$$

Equations (3.12), (3.15) and (3.16) imply eq. (3.6), while eqs. (3.11) and (3.15) imply eq. (3.5). Therefore, x_{ij}^{o} is a feasible solution of problem (3.4) - (3.7). Moreover, it is also optimal. In fact, observe that (x_{ij}^{o}, y_{ij}^{o}) satisfying eq. (3.15) is a feasible solution of problem (3.10) - (3.14) with the two extra constraints

$$y_{ij} = \frac{x_{ij}}{\sum_{i,j} t_{ij} x_{ij}} \qquad i,j = 1,\dots,n \tag{3.17}$$

$$x_{ij} = \text{non negative integer} \quad i,j = 1,\dots,n . \tag{3.18}$$

Hence (x^0_{ij} , y^0_{ij}) is optimal for problem (3.10)-(3.14), (3.17), (3.18), but this problem is equivalent to problem (3.4) - (3.7).

What stated above can be summarized in the following algorithm.

<u>Outline of the algorithm</u>

1. Determine y^0_{ij} by solving the Linear Program (3.10) - (3.14).
2. Put y^0_{ij} in the form (3.15).
3. Determine $\Gamma_1, \dots, \Gamma_r$ and $n_1, \dots, n_r$ by applying Procedure 1 to x^0_{ij} . ▨

<u>Example</u>

Consider the finite automation with $n = 3$ and $\bar{V} = 4$ described in Fig. 3.2, where the entries correspond to possible state transitions.

c_{ij}	1	2	3
1	6	1	
2	1	12	9
3	5		2

t_{ij}	1	2	3
1	2	1	
2	1	3	3
3	1		1

V_{ij}	1	2	3
1	2	2	
2	2	6	5
3	8		5

Fig. 3.2.: The costs C_{ij}, times t_{ij} and constraints V_{ij}.

The solution of the Linear Program (3.10) - (3.14) is given by

$$y^o_{12} = y^o_{21} = \frac{1}{6}$$

$$y^o_{33} = \frac{2}{3}$$

$$y^o_{ij} = 0 \quad \text{for all other} \quad (i,j).$$

Step 2 of the algorithm gives

$$y^o_{12} = \frac{x^o_{12}}{T} = \frac{1}{6}$$

$$y^o_{21} = \frac{x^o_{21}}{T} = \frac{1}{6}$$

$$y^o_{33} = \frac{x^o_{33}}{T} = \frac{4}{6} .$$

Thus Procedure 1 can be applied. The first iteration is

$$n_1 = 1$$

$$x^{(1)}_{i_1 i_2} = x^{(1)}_{12}$$

$$x^{(1)}_{i_2 i_3} = x^{(1)}_{21}$$

$$\text{i.e.,} \ \Gamma_1 = (1-2-1).$$

The second iteration is

$$x_{33}^{(2)} = x_{33}^{(1)} = 4$$

$$x_{12}^{(2)} = x_{21}^{(2)} = 0$$

$$n_2 = 4$$

$$x_{i_1 i_2}^{(2)} = x_{33}^{(2)}$$

$$\text{i.e.} \quad \Gamma_2 = (3-3).$$

Hence the solution of the algorithm is given by cycle Γ_1 repeated n_1 times and cycle Γ_2 repeated n_2 times. Since Γ_1 and Γ_2 do not have common nodes, they do not constitute a cycle Γ and the problem does not have an optimal solution. Neverthless, the value

$$C^0 = \frac{n_1 \mathscr{C}(\Gamma_1) + n_2 \mathscr{C}(\Gamma_2)}{n_1 \tau(\Gamma_1) + n_2 \tau(\Gamma_2)} = \frac{5}{3}$$

can be approximated at will by a cycle Γ constituted by Γ_1 run $n_1 k$ times, Γ_2 run $n_2 k$ times and $\Gamma_3 = (1\text{-}2\text{-}3\text{-}1)$ run only once. For instance, $k = 100$ gives

$$C(\Gamma) = \frac{1015}{605}$$

$$V(\Gamma) = \frac{2415}{605}.$$

4. Applications

Unconstrained Periodic Optimization problems arise in connection with a number of different applications such as transportation problems [6], PERT cost schemes [7], serving systems [7] and scheduling parallel computation [9].

Aim of this section is to provide two significant examples of constrained Periodic Optimization.

As for path constrained problems suppose that the best route of a vessel for hire has to be selected [6]. A profit r_{ij} is made each time the vessel goes from port i to port j, while the time spent is t_{ij}. Obvious human and technical factors cause the vessel to reach a particular port (say port 1) at least once every N units of time. The performance to be maximized is the profit per unit time. The problem can be readly formulated as a path constrained Periodic Optimization problem and solved applying the algorithm presented in Sect. 2.

An example of integral constrained Periodic Optimization can be derived from a particular production sequencing problem [10]. More precisely, the best cyclic operation of a plant for colouring artificial fibres must be found.

Let $k = 1, \dots, q$ be all possible colours. Due to the particular structure of the plant two fibres can be simultaneously processed, not necessarily with the same colour.

If $i = 1, \dots, n$ are the possible states of the plant (i.e. couple of colours), the problem can be described on a complete graph with n nodes. Let i_k be the node corresponding to treating both fibres with colour k, and X_k the set of nodes corresponding to treating only one of the fibres with colour k. Then, $2 + q$ weights can be associated in the following way to each arc (i, j) of the graph, provided that a suitable unit of time has been chosen.

$$t_{ij} = \begin{cases} \text{positive integer} & i \neq j \quad \text{(set-up time)} \\ 1 & i = j \end{cases}$$

$$r_{ij} = \begin{cases} 0 & i \neq j \quad \text{(no profit is made when setting-up)} \\ > 0 & i = j \end{cases}$$

$$v_{ij}^{(k)} = \begin{cases} 0 & i \neq j \quad \text{(no production at all when setting-up)} \\ 0 & i = j \notin X_k \cup i_k \\ v & i = j \in X_k \\ 2v & i = j = i_k \end{cases}$$

If $\bar{V}^{(k)}$, $k = 1, \dots, q$ is the minimum allowed mean production of fibre of colour k, the problem consists in finding the cycle of maximum mean profit which satisfies the production requirements. Thus, the problem can be handled by means of the algorithm presented in Section 3.

5. Concluding Remarks

The problem of Periodic Optimization of finite automata has been considered in this work.

Two kinds of constrained problems have been dealt with, namely the so-called "path constraint" and "integral constraints" problems.

The first one has been solved by taking into account the fact that the optimal cycle has the property of passing only once through the node which is constrained to be met.

As for the second problem the optimal solution is found by solving a standard Linear Program which gives the elementary cycles constituting the optimal cycle. Moreover, it has been proved that at least one optimal cycle can be decomposed in at most ($q + 1$) elementary cycles where q is the number of integral type constraints.

REFERENCES

[1] M. BELLMORE, G.L. NEMHAUSER, " The travelling Salesman Problem: a Survey" Operations Research, Vol 16, pp. 538-558, 1968.

[2] M. HELD, R.M. KARP, "The Travelling Salesman Problem and Minimum Spanning Trees" , Part I Operations Research, Vol. 18, pp. 1138-1162, 1970.

[3] M. HELD, R.M. KARP, " The Travelling Salesman Problem and Minimum Spanning Trees" Part II Mathematical Programming, vol. 1, pp. 6-25 1971.

[4] L. FRATTA, F. MAFFIOLI, "On Heuristically Guided Algorithm for the Travelling Salesman Problem" Proceedings of the Int. Congr. of Cybernetics and Systems, August 28-September 3, 1972, Oxford, U.K.

[5] E.L. LAWLER, " Optimal Cycles in Doubly Weighted Directed Linear Graphs" Theorie des Graphes, Int. Symp. Roma, Italy, July 1966, Dunod, Paris.

[6] G.B. DANTZIG, W.O. BLATTNER and M.R. RAO, " Finding a Cycle in a Graph with Minimum Cost to Time Ratio with Application to a Ship Routing Problem", Theorie des Graphes Int. Symp. Roma Italy, July 1966, Dunod, Paris.

[7] I.V. ROMANOVSKII, " Optimization of Stationary Control of a Discrete Deterministic Process" Kibernetika, Vol. 3, N.2, pp. 66-78, March-April 1967. (American translation in Cybernetics).

[8] B. FOX, " Finding Minimal Cost-Time Ratio Circuits"

Operations Research, vol. 17, N.3, pp. 564-550, May-June 1969.

[9] R. REITER, "Scheduling Parallel Computations", J. of ACM, vol. 15, N.4, pp. 590-599, Oct. 1968.

[10] P. MIGLIARESE, P. PALERMO, C. ROVEDA, "A Problem of Sequencing for Artificial Fibres Industries", Ricerca Operativa, vol.1, N.1, pp. 37-40, Jan. 1971, (in Italian).

FRANCESCO MAFFIOLI (*)

OPTIMAL HAMILTONIAN CYCLES : A SURVEY OF RECENT RESULTS (**)

(*) Istituto di Elettrotecnica ed Elettronica, Politecnico di Milano, Milano, Italy

(**) Partially supported by Centro di Telecomunicazioni Spaziali (CNR) of Milano, Italy

1. Introduction

The problem of finding a shortest elementary (*) cycle spanning all the nodes of a given arc weighted graph is called traveling salesman problem (TSP) and any solution is called a shortest hamiltonian cycle (SH) or a shortest tour of the given graph.

More precisely let $G(N, A)$ be the given graph, where $N = \{1, 2, \ldots, n\}$ and $A = \{(i, j)\}$ are respectively the set of nodes and the set of arcs of G. Assume also a real positive number c_{ij} is assigned for each arc (i, j) ($c_{ij} = \infty$ if $(i, j) \notin A$). A sequence of arcs of A is sought

$$S = \{(i_1, i_2), (i_2, i_3), \ldots, (i_n, i_1)\}$$

such that $(i_1, i_2, \ldots, i_n)$ represents a suitable permutation of the nodes of N and such that

$$z = \sum_{(i,j) \in S} c_{ij}$$

be minimum. $C = \{c_{ij}\}$ can be considered from time to time as a matrix either of lengths, or costs, or time delays, or algorithms of failure-probability of the corresponding arcs: the first of these meanings is used here just for the sake of sim

---

(+) That is passing through each node exactly once.

plicity.

The extremely large literature existing about the TSP points out sufficiently enough its importance. In fact the TSP arises as a key-step towards solving important problems, in a large variety of practical fields, such as scheduling problems, optimal control, reliable communication and transportation networks, etc.

A rather complete survey of works on the TSP up to 1968 has been published [1] , [2] : the purpose of the present work is to survey some recent results and present new approaches to the TSP.

Let us recall that a problem is considered solved from the point of view of computational complexity if there exists a method of solution for which a polynomial bound on the computation time as a function of the "size" of the problem is known. For instance the problem of finding a shortest spanning tree (SST) of a given graph is solved in this sense since a program can be written whose computing time is of the order of magnitude of $O(n^2)$, n being the number of nodes of the graph [14] , [16] . For the TSP the best known upper bounds are instead exponential functions of n , so that this problem may not be considered solved. This fact has suggested to develop several heuristic methods to obtain suboptimal solutions in a reasonable amount of computation time even for very large problems.

Some of the most recent ones are listed among the

references and will not be considered here [7], [13].

Heuristic methods are not to be confused with heuristically guided ones, which are exact and will occupy a large part of this work.

Two special cases are known to the author for which a polynomially bounded algorithm exists: one arises in connection with a problem of sequencing a one-state variable machine [3], while the other considers the case of C being upper triangular, i.e. of $c_{ij} = 0$ for all $i \geq j$ [4]. It can be shown that in this second case the TSP can be reduced to the optimum assignment problem, which is known to be solvable by a polynomially bounded algorithm of the order of n^3.

2. Some Related Problems

A problem strictly related to the TSP is that of finding the shortest (or the longest) elementary spanning path of a graph. For similarity with a hamiltonian cycle (H), an elementary spanning path is called a hamiltonian chain (HC).

Assume now that the TSP is assigned on a graph $G(N, A)$. A new graph $G'(N', A')$ may be constructed with $n' = n+1$ nodes (one more than G) and where the set of arcs A' contains all the arcs of A but for those going into, say, node 1, which in G' are directed towards the new node (see fig. 1). For what concerns the costs of the arcs put

$$c_{ij} = c'_{ij} \quad \text{for} \quad j \neq 1 \quad , n+1$$

and

$$c_{i,n+1} = c'_{i1} \ .$$

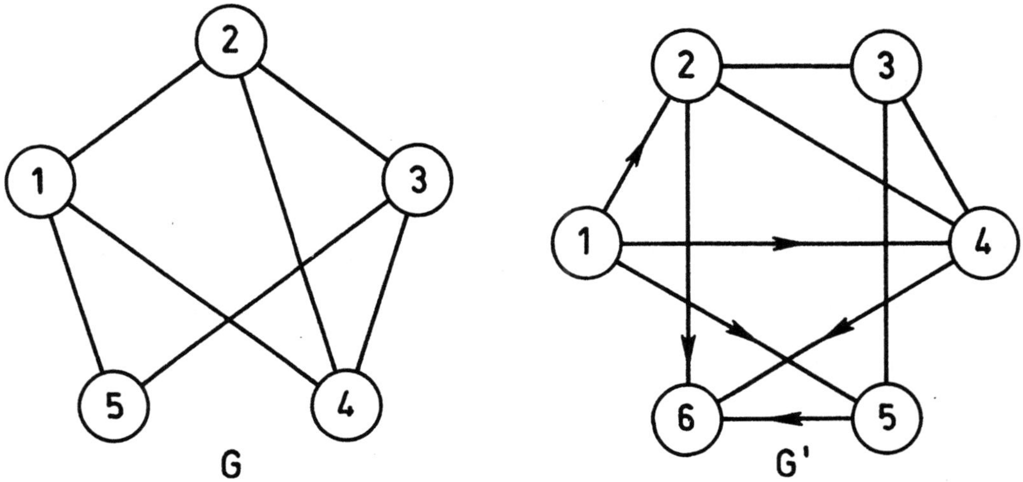

Fig. 1

It is well known that a one-to-one correspondence exists between the SHC of G' and the SH of G . In fact if the SHC of G' is characterized f.i. by the sequence of nodes

$$\{1,4,2,3,5,6\}$$

the SH on G is given by the sequence of nodes

$$\{1,4,2,3,5,1\}.$$

Since the amount of computation time which is necessary to obtain G' from G and the SH on G from the SHC on

G' is obviously very small, any algorithm which would solve the SHC problem would solve the TSP and viceversa. This approach to the TSP has been recently followed by several authors. In [8] two algorithms are proposed, based both on the followings theorems.

<u>Theorem 1:</u> a HC is a spanning tree (ST) such that no node has degree (*) greater than 2.

<u>Theorem 2</u>: If a SST is a HC then it is a SHC.

As a matter of fact the SHC problem was also considered in an earlier work [15] which presents a linear programming formulation of the problem. For a completely connected graph with n nodes, this formulation would involve $n(n+1)$ variables and $n(n+3)/2+1$ constraints to be considered explicitely together with a large number of constraints to be considered implicitely by introducing a few of them at a time after each iterative application of the simplex method. This approach is therefore considered complicate and inefficient from the authors themselves [15] .

Although much simpler, the algorithms presented in [8] do not prove to be very efficient either.

Algorithm A takes advandage of Theorems 1 and 2, implementing a branch-and-bound procedure, which starts from

---

(*) The degree of a nodes is equal to the number of arcs incident to that node.

the SST of G and attempts to find the HC whose total length differs as little as possible from the length of the SST. An example is shown in fig. 2.

The most important result of [8] is perhaps the following theorem on which algorithm B will be founded, and which shall be used also in the sequel.

<u>Theorem 3</u>: if the cost matrix C is transformed into a matrix C' so that

$$c'_{ij} = c_{ij} + \pi_i + \pi_j \quad \text{for all} \quad i,j = 1,2,\dots,n$$

where π is an n-dimensional vector of real positive or negative constants, then the relative costs under C' of all HC having specified end nodes remain unchanged.

<u>Proof</u>: [8] let c and c' be the costs under C and C' respectively of an arbitrary HC having end nodes i_1 and i_2.

Then

$$c - c' = 2 \sum_{j \neq i_1, i_2} \pi_j + \pi_{i_1} + \pi_{i_2}$$

which is a constant amount irrespective of the HC chosen. Q.E.D.

Algorithm B of [8] is a node-penalty method based upon Theorems 1, 2 and 3, and may be formalized as follows (where d_i is the degree of the i-th node):

Step 1: compute the SST of G under cost matrix C.

If $d_i \leq 2, \forall i$ stop, since a SHC has been found.

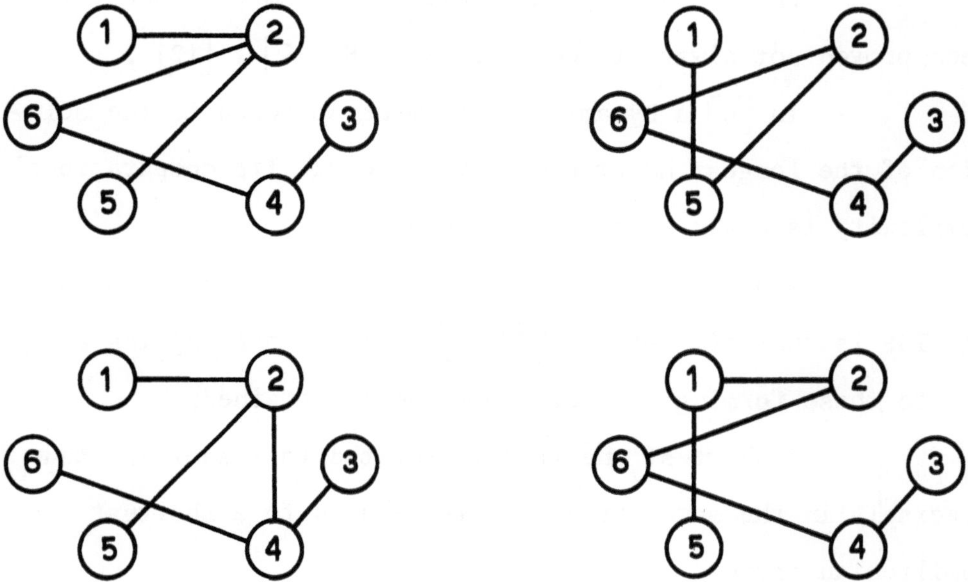

$$C = \{c_{ij}\} = \begin{bmatrix} - & 4 & 10 & 18 & 5 & 10 \\ 4 & - & 12 & 8 & 2 & 6 \\ 10 & 12 & - & 4 & 18 & 16 \\ 18 & 8 & 4 & - & 14 & 6 \\ 5 & 2 & 18 & 14 & - & 16 \\ 10 & 6 & 16 & 6 & 16 & - \end{bmatrix}$$

Fig. 2

Step 2: assign penalties π_i to the nodes of G in any way but positive if $d_i > 2$, zero if $d_i = 2$ and negative if $d_i = 1$, except for the initial and final nodes.

Step 3: compute a new cost matrix $C = \{\bar{c}_{ij}\}$ such that

$$\bar{c}_{ij} = c_{ij} + \pi_i + \pi_j \qquad i,j = 1,2,\dots,n$$

Go to step 1.

Unfortunately, the above algorithm has recently been proved not always to converge at a HC [5] , [12] .

In [9] a Dynamic Programming Approach to the solution of the longest HC problem may be found. Its computational complexity is again exponentially bounded.

Another concept introduced recently to approach the TSP is that of 1 -trees [5] , [6] . When one adjoins an arc to those forming a tree, a 1 -tree is obtained.

A shortest one-tree is related to a shortest tour precisely in the same way a tree is related to a shortest hamiltonian chain.

In [5] Held and Karp have tried to utilize the concept of 1 -trees to develop an efficient algorithm for the TSP. They succeeded only in part and recently [6] a branch-and-bound procedure which has to be considered the most efficient method for solving the TSP experimented up to day appeared.

Most of their results have been embedded with some previous work of Fratta and the present author [11] to yield what is believed to be a very promising approach [12] now under advanced experimentation. In what follows the main results of [5] , [6] and [12] will be summarized.

3. Linear Programming Approach[5]

The following properties of 1 -trees and tours are easily proved as they are or letting derive them from the corresponding previously mentioned properties of trees and HC, recalled between brackets.

i) A tour is a 1 -tree in which each node has degree 2 (Theorem 1).

ii) If a minimum-cost 1 -tree is a tour, then it is a tour of minimum cost (Theorem 2).

iii) For any real n -vector $\underset{\sim}{\pi} = (\pi_1, \pi_2, \dots, \pi_n)$, the transformation on arc costs

$$c_{ij} \rightarrow c_{ij} + \pi_i + \pi_j$$

leaves the TSP invariant, but changes the minimum 1-tree (Theorem 3).

Clearly

$$c^* + 2 \sum_{i=1}^{n} \pi_i \geq \min_k \left[c_k + \sum_{i=1}^{n} \pi_i d_{ik} \right] \qquad (3.1)$$

where c^* is the cost of a minimum tour with respect to $C = \{c_{ij}\}$, k indexes the 1 -trees, c_k is the cost of the k -th 1 -tree with respect to C and d_{i_k} is the degree of node i in the k -th 1 -tree. If

$$w(\underset{\sim}{\pi}) = \min_k \left[c_k + \underset{\sim}{\pi} \cdot \underset{\sim}{\nu}_k \right]$$

where $\underset{\sim}{\nu}_k$ is a vector having $d_{ik}-2$ as its i-th component and $\cdot$ denotes inner product, then (3.1) becomes

$$c^* \geq w(\underset{\sim}{\pi}) \tag{3.2}$$

and one obtains an infinite family of lower bounds on the cost of an optimum tour, the best such bound being $\max_{\underset{\sim}{\pi}} w(\underset{\sim}{\pi})$.

Unfortunately there are cases in which the equal sign in (3.2) is never obtainable as it is proved in [5]. Nevertheless finding the vector $\underset{\sim}{\pi}$ for which $w(\underset{\sim}{\pi})$ reaches its maximum could be very helpful.

The problem of finding

$$\max_{\underset{\sim}{\pi}} w(\underset{\sim}{\pi})$$

is equivalent to the following linear programming problem

$$\left\{ \max w : w \leq c_k + \underset{\sim}{\pi} \cdot \underset{\sim}{\nu}_k , \quad \forall k \right\}$$

which, dualizing, becomes

$$\min \sum_k y_k c_k : \sum_k y_k = 1 , \; y_k \geq 0 , \; \sum_k \nu_{ki} y_k = 0 \quad i = 2,3,\dots,n-1 \tag{3.3}$$

Following (3.3) we may calculate $\max_{\pi} w(\underset{\sim}{\pi})$ and a Branch-and-Bound procedure has been developed, which uses this technique to calculate the bounds. Unfortunately, this has been proved effective only up to 20 or 30 nodes, since the convergence of linear programming method is quite slow [5].

4. Ascent Method Approach[6]

An iterative method for computing or approximating $\max_{\pi} w(\pi)$ may be based on the following sequence of n-vectors $\{\pi^m\}$

$$\pi^{m+1} = \pi^m + t_m v_k(\pi^m) \qquad (4.1)$$

where for any π, $k(\pi)$ is the index of a minimum-cost 1-tree at the point π and $\{t_m\}$ is a sequence of scalars.

In [6] the following lemmas are proved.

Lemma 1: let $\bar{\pi}$ and π be such that $w(\bar{\pi}) \geq w(\pi)$. Then

$$(\bar{\pi} - \pi) \cdot v_{k(\pi)} \geq w(\bar{\pi}) - w(\pi) \geq 0.$$

Lemma 2: (appropriate step sizes): if

$$0 < t < \frac{2(w(\bar{\pi}) - w(\pi))}{\| v_{k(\pi)} \|^2}$$

then

$$\| \bar{\pi} - (\pi + t) v_{k(\pi)} \| < \| \bar{\pi} - \pi \| ^{*)}.$$

Even the simple strategy of taking $t_m = \bar{t}$, $\forall m$ proves effective, because of the following

---

(*) $\| \cdot \|$ denotes Euclidean norms.

Theorem 4 (theorem 1 in [6]): the iteration (4.1) with $t_m = \bar{t}$, $\forall\, m$, satisfies

$$\sup_m w(\underset{\sim}{\pi}^m) \geq \max_{\pi} w(\underset{\sim}{\pi}) - \frac{1}{2}\bar{t} \limsup_{m\to\infty} \| \underset{\sim}{\nu}_{k(\underset{\sim}{\pi}^m)} \|^2 .$$

This is no wonder since $\underset{\sim}{\nu}_{k(\underset{\sim}{\pi})}$ is nothing less than the gradient of $w(\underset{\sim}{\pi})$ and the iteration scheme (4.1) is therefore equivalent to a true ascent search.

It is also immediate to see now why algorithm B of [8] is effective but not always convergent. Translating in fact what has seen for 1 -trees to the case of HC's , algorithm B is nothing more than the iteration scheme (4.1) which even when it reaches the optimum may fail to yield a HC in [8] as it may fail to yield a tour in [6] .

The branch-and bound procedure developed in [6] from the previously seen arguments is more or less standard and will not be reported here.

Computational results have been highly succesful: in fact it is possible for the authors something which has never been previously possible, that is to present entirely the search trees for TSP up to $n = 64$ [6] , in corresponding figures.

5. Heuristically Guided Search [10]

This algorithm is used in the last approach to the

TSP we shall review here and is therefore shortly summarized for the reader's convenience.

The algorithm finds a minimal path in a weighted decisiongraph, whose vertices are the different " states" of the problem.

Although developed independently by Hart, Nilsson and Raphael [10] this algorithm may be viewed quite simply as a modification of the famous Moore-Dijkstra-Dantzig algorithm [17].

Let $D(V,T)$ be a finite, directed graph of vertices $V_i \in V$ and links $(V_i, V_j) \in T$; to each lind a weight t_{ij} is attached.

A sequence of links $(V_{i_1}, V_{i_2}), (V_{i_2}, V_{i_3})$; ..., $(V_{i_{r-1}}, V_{i_r})$ is a path between V_{i_1} and V_{i_r} and $d_{1,r} = \sum_{k=1}^{r-1} t_{i_k, i_{k+1}}$ is its cost.

Given an initial vertex V_a and a final vertex V_z the shortest path problem consists in determining a path from V_a to V_z with minimal cost. In what follows the cost $d_{i,j}$ of the minimal path between vertex V_i and V_j will be called the <u>distance</u> between V_i and V_j .

Starting from the vertices adjacent to V_a , a parameter p_i is defined for every vertex V_i .

In any intermediate step the vertices are partioned with respect to p_i into three categories:

a) Closed vertices, for which p_i is known to be the distance from V_a : $p_i = d_{a,i} = d_i$.

b) Open vertices, for which $p_i \geq d_i$.

c) Blank vertices, for which a value of p_i has not yet been computed.

Assume further that an estimate q_i of the distance $e_i = d_{i,z}$ between V_i and V_z is available from the problem domain.

A <u>total estimate</u> for node V_i is defined as $f_i = p_i + q_i$. Moreover assume that a triangular property holds, that is

(5.1) $$q_i \leq d_{i,j} + q_j ,$$

which implies that q_i must be a lower bound to e_i. Inequality (5.1) is called the <u>consistency assumption</u>.

A typical step of the algorithm consists in <u>closing</u> the open vertex V_i with minimum f_i and in <u>expanding</u> V_i, i.e., computing again first the p_k's of all open or blank vertices V_k such that $(V_i, V_k) \in T$ by the formulas

(5.2) $$\bar{p}_k = p_i + t_{ik} \qquad \text{if } V_k \text{ was blank}$$

$$\bar{p}_k = \min(p_k, p_i + t_{ik}) \qquad \text{if } V_k \text{ was open}$$

and then corresponding q_k's and f_k's.

The algorithm follows.

<u>Step 1.</u> Open the initial node V_a with $p_a = 0$, $q_a = f_a$.

<u>Step 2.</u> Close the open vertex V_i for which the total estimate

$$f_i = p_i + q_i$$

is minimal. If more than one vertex satisfies this condition and V_z is one of them, choose V_z, otherwise choose any one of them. Thus $p_i = d_i$.

Step 3. Stop if the goal node was closed in step 2.

Step 4. Expand V_i. Go to step 2.

For large graphs the entire graph is not stored in the memory, but rules for generating adjacent vertices are given: only when V_i is not blank its name, parameters p_i and f_i have to be stored.

The above algorithm is strictly related both to Branch-and-Bound algorithms and to Dynamic Programming, having some advantages of both these methods. With respect to Branch-and-Bound methods it has the advantage of never considering as two distinct vertices of the decision graph the same state of the problem.

The efficiency of the heuristically guided search is of course highly dependent upon how close the estimate is to the real value of the minimal solution.

It can be shown [10] that if the estimate q_i is exact, i.e., if $q_i = e_i$ for all V_i, the algorithm expands only vertices on optimal paths between V_a and V_z.

6. Heuristically Guided Algorithm for the TSP[12]

As previously seen any algorithm which wuold find the SHC of a graph would also solve the TSP and viceversa. The approach of [12] begins therefore constructing a new graph whose

SHC's would be in direct correspondence with the shortest tours of the original graph.

The main part of the algorithm is devoted to built up a SHC starting from one node, say n_1.

This fact does not imply any lack of generality since by adding to G a new node n_0 connected to all the others by arcs of unitary weight the problem of finding the SHC of G irrespectively of the starting node is reduced to the problem of finding the SHC of the new graph starting from n_0.

In order to apply the general concepts reviewed in the previous section to our problem we have to show first of all how a decisional graph $D(V,T)$ may be derived. Each vertex V_i is a <u>state</u> of the problem defined as a subset of nodes of G connected by a HC which starts from n_1 and terminates at node n_k. The solution vertex V_z represents a state corresponding to N, the set of all nodes of G. The vertex V_a corresponds to a subset containing only the node n_1.

The weight t_{ij} of the link connecting V_i and V_j in D is equal to the weight $c_{k,h}$ of the arc of G connecting the last node n_k of V_i to the node n_h added to the set of nodes characterizing V_i to obtain the nextstate V_j.

The estimate p_i assigned to V_i is given by the sum of the weights of the links of the path connecting V_a to V_i. As soon as V_i is closed, p_i corresponds to the SHC on the subset of nodes characterizing the state V_i, that is to the distance $d_{1,i} = d_i$ as defined in the previous section.

As estimate q_i of the distance $d_{i,z}$, i.e., of the remaining part of the SHC, let us assume as a first possibility the cost of the SST connecting n_k to all the nodes which do not belong to the state V_i .

A well known property of the shortest spanning tree is the following: given a set of nodes X divided into two subsets X_1 and X_2 having just one node in common, the cost of the SST connecting all nodes of X is less than or equal to the sum of the costs of the two SST-s connecting respectively all nodes of X_1 and all nodes of X_2 .

We are now in position to prove that the consistency assumption is valid in our case, i.e., that

$$q_i \leq d_{i,j} + q_j \tag{6.1}$$

By the above mentioned property we have

$$q_i \leq c + q_j \tag{6.2}$$

where c is the cost of the SST connecting node n_i to n_j and to all nodes which, in the sequence S characterizing the SHC, are between n_i and n_j .

From Theorem 2 we have the corollary:

$$c \leq d_{i,j}$$

and therefore from (6.2),

$$q_i \leq d_{i,j} + q_j.$$

Q.D.E.

Let us note that the algorithm for computing the SST of a set of nodes is among the most efficient ones [16] and therefore it should cause no harm having to use it even many times. Moreover in practice there are many $V_i^{'}s$ for which the estimate is the same.

It was pointed out in the previous paragraph that using heuristically guided search method is much more rewarding if the estimate q_i is as good as possible.

Using as estimate the SST as we have done up to now is by no means the best trying which can be done. In fact as was already pointed out we can use for estimating SCH's pretty much the same considerations used in [6] for estimating optimum tours. To be more precise let us introduce some notations.

Let c_0 be the cost of a SHC connecting a set of nodes $X = \{1,2,\ldots,p\}$, 1 and p being respectively the initial and the final node, with $C = \{c_{ij}\}$ as cost matrix.

Then under the cost matrix

$$C' = \{c'_{ij}\} = \{c_{ij} + \pi_i + \pi_j\},$$

$\underset{\sim}{\pi}$ being a real valued p-dimensional vector, the new cost of the same SHC shall be

$$c_0 + 2 \sum_{2}^{p-1} \pi_i + \pi_1 + \pi_p .$$

Let c_k be the cost of the k-th tree T_k connecting the nodes of X under C. Then (d_{ik} = degree of i-th node of X in T_k)

$$c_k + \sum_{i=1}^{p} \pi_i \cdot d_{ik}$$

shall be the cost of the same tree under matrix C'.

Let us observe that while relative costs of HC's are unaffected by $\underset{\sim}{\pi}$, this cannot be said for trees, precisely in the same way $\underset{\sim}{\pi}$ did affect 1-trees', but not tours' relative costs.

Let now X_1 and X_2 be two subsets of X covering X and having only one node, say ℓ, in common:

$$X_1 \cup X_2 = X \qquad X_1 \cap X_2 = \{\ell\} .$$

Let c_k^1 (c_k^2) be the cost under C of T_k^1 (T_k^2) spanning X_1 (X_2). Similarly let c_0' be the cost of a SHC of X_1 under C and

$$c_0' + 2 \sum_{i=2}^{\ell-1} \pi_i + \pi_1 + \pi_\ell$$

the cost of a SHC of X under C^1.

From previous theorems and properties we have therefore:

(6.3) $$c_0 \geq \min_k c_k \text{ *)}$$

and

(6.4) $$\min_k c_k \leq \min_k c_k^1 + \min_k c_k^2 .$$

From (6.3) we have

$$c_0 + 2 \sum_{i=2}^{p-1} \pi_i + \pi_1 + \pi_p \geq \min_k \left[c_k + \sum_{i=1}^{p} \pi_i \cdot d_{ik} \right].$$

Which may be written as

$$c_0 \geq \min_k \left\{ c_k + \underset{\sim}{\pi} \cdot \underset{\sim}{\mu}_k \right\}$$

where

$$\mu_{ki} = \begin{cases} d_{ik} - 1 & \text{if } i = 1 \text{ or } p \\ d_{ik} - 2 & \text{ortherwise} \end{cases}$$

Let by definition

$$v(\underset{\sim}{\pi}) = \min_k \left[c_k + \underset{\sim}{\pi} \cdot \underset{\sim}{\mu}_k \right],$$

then

(6.5) $$c_0 \geq \max_{\underset{\sim}{\pi}} v(\underset{\sim}{\pi}) .$$

(*) $\min_k c_k$ is of course a SST, since k indexes the trees.

This suggests that a much better estimate q_i could be obtained using $v(\underset{\sim}{\pi})$ and calculating or approximating its maximum. Since $\nabla v = \underset{\sim}{\mu}_k$ an iteration scheme can be implemented here too, substituting $\underset{\sim}{\mu}_k$ to $\underset{\sim}{\nu}_k$ in (4.1). The following theorem is proved in [12] and establishes the applicability of this new estimate for an heuristically guided search method.

Theorem 5: let $\underset{\sim}{\pi}^o$ be the p-dimensional vector for which the maximum of $v(\underset{\sim}{\pi})$ is reached. Let $q_i = v(\underset{\sim}{\pi}^o)$. Then

$$v(\underset{\sim}{\pi}^o) \leq c_0^1 + \max_{\underset{\sim}{\pi}} v_2(\underset{\sim}{\pi})$$

where $v_2(\underset{\sim}{\pi})$ is defined over X_2 and may be therefore assumed as q_j, while c_0^1 is by definition d_{ij}. This proves the consistency assumption with these new estimate validating it for the heuristically guided search algorithm.

In practice $\underset{\sim}{\pi}^o$ may be reached only very slowly with (4.1) and the iteration has to stop somehow in between the cost of a simple SST and $v(\underset{\sim}{\pi}^o)$.

Then the consistency assumption is no longer valid and the algorithm has to consider this fact.

As it is shown in [10] the heuristically guided search still converges provided no state is closed permanently.

This algorithm is now being programmed.

7. Conclusions

The TSP may still be considered unsolved since no polynomially bounded algorithm has been found. This fact is considered by several people as intrinsic to this as to other problems (f.i., Steiner problem, covering, etc.) and it may well be that for these problems the only thing to do is to construct exponentially bounded algorithms which would work very well in the average. The algorithms reviewed here are approaches in this direction. Among them the heuristically guided method seems the most promising one, although a definite programming experience is not yet available.

8. Acknowledgment

It is a pleasure to aknowledge here the many fruitful discussions with L. Fratta and G. Tartara as well as the kind invitation of A. Marzollo of CISM to give this seminar.

REFERENCES

[1] M. Bellmore & G.L. Nemhauser, "The traveling salesman problem: a survey' Oper. Res. 16, N°3 (1968) 538-558.

[2] A.M. Isaac & E. Turban, "Some comments on the traveling salesman problem", Oper. Res. 17, N°3 (1969) 543-546.

[3] P.C. Gilmore & R.E. Gomory, "Sequencing a one-state variable machine: a solvable case of the traveling salesman problem' Oper. Res. 12 (1964) 655-679.

[4] E.L. Lawler, "A solvable case of the traveling salesman problem" Math. Progr. 1, N° 2 (1971) 267-269.

[5] M. Held & R.M. Karp, "The traveling salesman problem and minimum spanning trees", Oper. Res. 18 (1970).

[6] M. Held & R.M. Karp, "The traveling salesman problem and minimum spanning tree: part II, Math. Progr. 1 (1971), 6-25.

[7] P. Krolak & W. Felts, "A man-machine approach toward solving the traveling salesman problem" Comm. ACM 14, N° 5 (1971) 327-334.

[8] N. Chirstofides, "The shortest Hamiltonian Chain of a Graph", SIAM J. Appl. Math. 19, N° 4 (1970) 689-696.

[9] A.G. Azpeitia & D. Riley, " The longest path in a network" , J. Math. An. & Appl. 30 (1970) 636-638.

[10] P.E. Hart, N.L. Nillson & B. Raphael, " A formal basis

for the heuristic determination of minimum cost paths" IEEE Trans. on SSC,SSC-4 N° 2 (1968) 100-107.

[11] L. Fratta & F. Maffioli, " On the shortest Hamiltonian chain of a network', 2nd International Symp. on Network Theory, Herceg-Novi (YU) July 1972.

[12] P.M. Camerini, L. Fratta, F. Maffioli, "A Heuristically guided Algorithm for the Travelling Salesman Problem", (to be published).

[13] K. Steiglits & P. Weiner, " Some improved algorithms for computer solution of the traveling salesman problem" Proc. 6th Allerton Conference on Circuits and System Theory (1968) 814-821.

[14] E.L. Lawler, " The complexity of combinatorial computations: a survey" , Proc. 1971 Polytechnic Institute of Brooklyn Symp. on Computers & Automata.

[15] N. Deo & S.L. Hakimi, " The shortest generalized Hamiltonian tree" Proc. 3rd Allerton Conference on Circuits & System Theory (1965), 879-883.

[16] V. Kevin & M. Withney, " Algorithms 422: Minimal Spanning tree [H] " , Comm. of ACM vol. 15, N° 4 (1972) 273-274.

[17] G.B. Dantzig, "Linear Programming and Extensions", Princeton U. Press 1963, 361-366.

SERGIO BITTANTI

GIORGIO FRONZA

GUIDO GUARDABASSI (*)

OPTIMALITY CONDITIONS FOR DISCRETE PERIODIC OPTIMIZATION PROBLEMS (**)

(*) Istituto di Elettrotecnica ed Elettronica, Politecnico di Milano, Milano, Italy

(**) Supported by Centro di Teoria dei Sistemi (CNR)

1. Problem Statement

A discrete periodic optimization problem can be stated in the following way. Consider a time invariant discrete system described by a set of difference equations of the form

$$x(k+1) = f\big(x(k),\, u(k)\big) \qquad k = 0,1,\ldots,N-1 \tag{1.1}$$

where

$$\begin{aligned} u(k) &\in U \subset R^m && \text{control vector} \\ x(k) &\in X = R^n && \text{state vector} \end{aligned} \tag{1.2}$$

and $f(\cdot,\cdot)$ is continuous together with its derivatives in $X \times U$. Then, maximize an average performance index

$$\mathcal{J}\left[u(\cdot);N\right] \triangleq \frac{1}{N}\sum_{k=0}^{N-1} g\big(x(k),u(k)\big) \tag{1.3}$$

subject to the periodicity constraint

$$x(0) = x(N)\,. \tag{1.4}$$

The function $g(\cdot,\cdot)$ is also assumed to be continuous together with its derivatives in

Three approaches to Problem (1.1)-(1.4), i.e. Dynamic Programming, Mathematical Programming and mainly the Maximum Principle will be discussed here.

2. The Dynamic Programming Approach

First we assume that N has a fixed value $\bar{N}$. Then system (1.1) subject to constraint (1.4) can be described by means of the cyclic block diagram of Fig. 1 and the optimization procedure introduced in [1] can be applied.

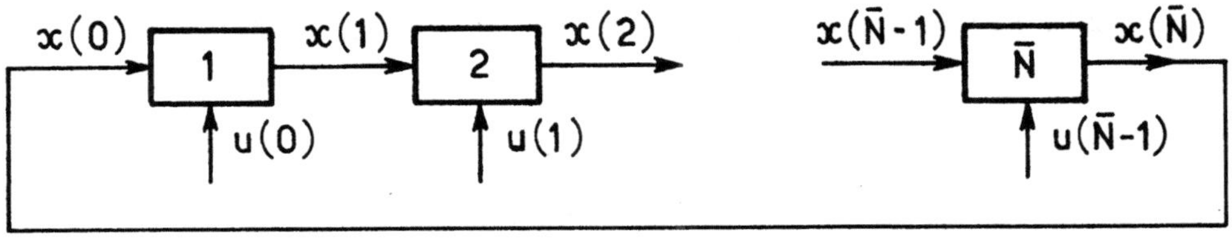

Fig. 1

In fact, problem (1.1)–(1.4) can be reformulated through the following steps.

i) Remove the loop identity by cutting the feedback line in the scheme of Fig 1; hence the cyclic system is converted into a serial one.

ii) Give the cut state $x(\bar{N})$ a fixed value c.

iii) Consider the final transition equation

$$c = f\left(x(\bar{N}-1)\ ,\ u(\bar{N}-1)\right)$$

and assume that it is possible to perform a decision inversion, i.e. to find $f^*(\cdot,\cdot)$ such that

(2.1) $$u(\bar{N}-1) = f^*(x(\bar{N}-1), c).$$

iv) By applying (1.1), (1.4), (1.5) convert the performance index (1.3) into a form $\hat{\mathfrak{J}}\left[c, u(0), u(1), \ldots, u(\bar{N}-2)\right]$

v) Solve the $\bar{N}$ decision-no state optimization problem

$$\max_{\substack{c \in X \\ u(1), u(2), \ldots, u(\bar{N}-2) \in U}} \hat{\mathfrak{J}}\left[c, u(0), u(1), \ldots, u(\bar{N}-2)\right].$$

Such a problem may be decomposed into $(\bar{N}-2)$ one decision-one state and 1 one decision-no state optimization problems by making use of the standard technique based on Dynamic Programming. By letting c take different values it is possible to build up a function $\varphi(c)$ which has to undergo a maximization with respect to c.

Finally, once a solution has been found for $N=\bar{N}$, through a slight additional effort it is possible to obtain an optimal solution for all $N<\bar{N}$.

It has been proved in [2] that, within the framework of Dynamic Programming, the method proposed in [1] and recalled above can be rated as an appealing one from the memory requirement point of view.

3. The Mathematical Programming Approach

As it is known, many optimal control problems have been solved in the past few years by following a Mathematical Programming Approach (see [3] for example) and the experience in the field seems to point out this method as the best one for facing large scale problems. It must be noted, however, that in this case the knowledge of the optimal solution to the mathematical program (1.1)-(1.4) for $N = \bar{N}$ does not supply any relevant information about the optimal solution for $N < \bar{N}$.

4. A Short Review of the Discrete Maximum Principle

Several attempts of extending Pontiyagin's Maximum Principle [4] to discrete optimal control problems have been made in the literature. A Discrete Maximum Principle was first devised by Katz in [5] and his results were developed by Fan and Wang [6], who also considered a problem characterized by a periodicity constraint. A direct version of Pontryagin's condition in the discrete case should be stated as follows. If $u^\circ(\cdot), x^\circ(\cdot)$, is an optimal solution and $\lambda^\circ(\cdot)$ is the corresponding co-state, then the Hamiltonian function must satisfy the inequality

$$H\left(x^\circ(k), \lambda^\circ(k+1), u\right) \leq H\left(x^\circ(k), \lambda^\circ(k+1), u^\circ(k)\right) \qquad k = 0,1,\ldots,N-1$$

for all $u \in U$.

As it was shown by Butkovskii in [7], this condition is not correct, since the Maximum Principle, stated under Pontryagin-like assumptions, has only a local validity in the discrete case. It is of fundamental importance, however, to remark that Butkovskii Local Maximum Principle has been stated only for the fixed initial state-free final state optimal control problem. In the general case it is not true even that the Hamiltonian must have a local maximum at the optimal control. In order to check the validity of this conclusion, consider the following elementary example. Let the system be described by

$$x(k+1) = -\frac{1}{2}x(k) + 3u(k) \qquad k = 0,1$$

under the constraint $x(0)=x(2)$, and take the performance index

$$J = \sum_{k=0}^{1} 3u^2(k) + 2u(k) - x^2(k) .$$

in this case the Hamiltonian is given by

$$H(x(k),\lambda(k+1),u(k)) = 3u^2(k) + 2u(k) - x^2(k) + \lambda(k+1)\left(-\frac{1}{2}x(k) + 3u(k)\right)$$

so that the adjoint system turns out to be

$$\lambda(k) = -\frac{1}{2}\lambda(k+1) - 2x(k) .$$

Through a direct variational analysis it is possible to check that the optimal constant solution $u^0(k)=1$, $x^0(k) = 2$, $\lambda^0(k) = -\frac{8}{3}$ is the general optimal solution too. In fact consider the control variation $\delta u(\cdot)$ defined by $\delta u(0)=\varepsilon, \delta u(1)=\alpha\varepsilon$ and the corresponding 2-periodic state variation $\delta x(\cdot)$ given by $\delta x(0)=2(2\alpha-1)\varepsilon, \delta x(1)=2(2-\alpha)\varepsilon$; it turns out that $\delta J=\varepsilon^2(-17+32\alpha-17\alpha^2)$ is negative for all ε and α. Moreover

$$H(x^0(k),\lambda^0(k+1),u) = 3u^2 - 6u + \frac{4}{3}$$

achieves a minimum at the optimal control.

Hence the conclusions is that no Discrete Maximum Principle, either in the large or local, can be formulated in the general case, at least without making further assumptions. In fact, Propoi [8], who applies Pontryagin's philosophy (i.e. the well-known "perturbation approach") to the discrete case and whose condition on the Hamiltonian is therefore independent of the particular boundary constraints (initial and final state-constraint set), states the following result: the function $H\left(x^{\circ}(k),\ \lambda^{\circ}(k+1),\cdot\right)$ must exhibit either a local maximum or a stationary value at the optimal control, (note that the latter occurs in the example above). This is the only weak condition of general validity for the discrete optimal control problem. It is not, however, the most original result supplied in 8. In fact by adding the basic hypothesis that the set

$$R(a)=\left\{\begin{vmatrix} x \\ x_0 \end{vmatrix} : \begin{matrix} x = f(a,u) \\ x_0 \leq g(a,u) \end{matrix},\ u \in U\right\}$$

be a convex subset of R^{n+1} for all $a \in R^n$, Propoi showed that Pontiygin's maximum condition can be stated in the discrete case with respect to all $u \in U$. This result does essentially coincide with the one independently supplied by Holtzmann and Halkin [9], [10] through the introduction of the concept of directional convexity. If z is a vector in R^q, a set $\mathcal{S} \subset R^q$ is said to be z-directionally convex if, for all $x, y \in \mathcal{S}$ and each $\mu \in [0,1]$, there exists $\Theta \geq 0$ such that $\mu x+(1-\mu)y+\Theta z \in \mathcal{S}$. Letting $e_{n+1}=\begin{bmatrix} 0_n \\ 1 \end{bmatrix}$, where 0_n is the zero n-vector, the convexity of $R(a)$ is equivalent to the e_{n+1}-

-directional convexity of the set

$$\tilde{R}(a) = \left\{ \begin{vmatrix} x \\ x_0 \end{vmatrix} : \begin{matrix} x = f(a,u) \\ x_0 = g(a,u) \end{matrix} \quad u \in U \right\}$$

According to Holtzmann and Halkin's terminology, the set $R(a)$ may be called the e_{n+1} -shadow of $\tilde{R}(a)$. It is worthwhile mentioning that the conditions found in [9] , [10] can also be derived as a particular application from the ones given by Canon, Cullum and Polak [11] for a very generally formulated optimization problem (Basic Problem). The impossibility of extending Pontryagin's condition to discrete systems was also pointed out by Gabasov and Kirillova [12] , [13] in a way completely different from the one followed in [7] - [10] . To be more specific, they discussed a discrete optimal control problem with the sampling period as a parameter, i.e. they considered the system

$$x(k+h) - x(k) = h\tilde{f}\,(x(k), u(k)) , \qquad h > 0$$

and found that, for any $\varepsilon > 0$, there must exist a sampling period $\bar{h}(\varepsilon)$ such that (Quasimaximum Principle)

$$H(x^\circ(k), \lambda^\circ(k+1), u) - \varepsilon \le H\,(x^\circ(k), \lambda^\circ(k+1), u^\circ(k)) ,$$

for $k = 0, h, \ldots, (N-1)h$, $h < \bar{h}(\varepsilon)$ for all $u \in U$.

On the contrary, the existence of such $\bar{h}$ is not guaranteed in correspondence with $\varepsilon = 0$, i.e. there are systems whose optimal control does not satisfy the maximum condition for any $h > 0$. This is, for instance, the case of the optimal control problem for the scalar system

$$x((k+1)h) = x(kh) + hu(kh) \qquad k=0,1$$

$$x(0) = 0$$

$$U = [-1,1]$$

and performance index

$$\mathcal{J} = \sum_{k=0}^{1} h\left[x^2(kh) - u^2(kh)\right].$$

Finally, in later works, [14] , [15] , Gabasov, Kirillova and Tarashenko obtained other necessary optimality conditions, i.e. they found high order conditions which must be satisfied by the Hamiltonian in a suitable finite region around the optimal control.

5. The Maximum Principle Approach to Discrete Periodic Optimization

Among the many different necessity optimality conditions cited above, here reference will be made only to the work [8] by Propoi (or to equivalent results [9] , [10]). This is due to the basic reason that the optimal periodic control problem (1.1)-(1.4) can be reformulated as an initial state constraint set-final state constraint set problem. Hence Propoi's conditions, based on a Pontryagin-type proof, can be directly applied. It must be stressed, however, that through all the following considerations, the period N will be considered as fixed.

Let

$$H(x(k),\lambda(k+1),u(k)) = g(x(k),u(k)) + \lambda'(k+1)f(x(k),u(k))$$

be the Hamiltonian for problem (1.1)-(1.4); moreover if $u^o(\cdot)$, $x^o(\cdot)$ is an optimal solution and $\lambda(\cdot)$ is the corresponding, co-state, let

$$A(k) \triangleq f_x(x^o(k),u^o(k)), \quad B(k) \triangleq f_u(x^o(k),u^o(k))$$

$$\Phi(k,j) \triangleq \prod_{m=1}^{k-j} A(k-m)$$

$$P(k) \triangleq H_{xx}(x^o(k),\lambda^o(k+1),u^o(k)),$$

$$Q(k) \triangleq H_{xu}(x^o(k),\lambda^o(k+1),u^o(k)), \quad R(k) \triangleq H_{uu}(x^o(k),\lambda^o(k+1),u^o(k))$$

Then the following optimality condition can be stated for problem (1.1)-(1.4).

<u>Theorem 1.</u> (Weak Pseudo-Maximum Principle)

If $u^o(\cdot), x^o(\cdot)$ is an optimal solution, then there must exist $\lambda^o(\cdot)$ such that

$$\delta^1 H(x^o(k),\lambda^o(k+1),u^o(k)) \leq 0 \qquad k=0,1,\ldots,N-1$$

for all admissible weak variations $\delta u(k)$ around the optimal control $u^o(k)$ and

$$\lambda^{o}(k) = H'_{x}(x^{o}(k), \lambda^{o}(k+1), u^{o}(k))$$

$$\lambda^{o}(N) = \lambda^{o}(0) .$$

Moreover if $\delta^{1}H(x^{o}(k), \lambda^{o}(k+1), u^{o}(k)) < 0$, then $u^{o}(k)$ is a boundary point of U ; if $u^{o}(k)$ is an internal point of U , and $\Phi(k+N+1, k+1)$ has no eigenvalue equal to 1, then

$$\delta^{1}H(x^{o}(k), \lambda^{o}(k+1), u^{o}(k)) = 0$$

and

$$R(k) + \Gamma'(k,N)\tilde{Q}(k,N) + \tilde{Q}'(k,N)\Gamma(k,N) + \Gamma'(k,N)\tilde{P}(k+1,k+N)\Gamma(k,N) \leq 0$$

where

$$\Gamma(k,N) = [I - \Phi(k+N+1, k+1)]^{-1} B(k)$$

$$\tilde{Q}(k,N) = \Phi'(k+N, k+1)\, Q(k)$$

and, for any $k \leq j$

$$\tilde{P}(k-1, j) = A'(k-1)\tilde{P}(k,j)A(k-1) + P(k-1)$$

$$\tilde{P}(j,j) = P(j)$$

Proof

First reformulate problem (1.1)-(1.4) in the following way.

Maximize

$$\hat{J} = z(N) \tag{5.1}$$

subject to

$$x(k+1) = f(x(k), u(k)) \tag{5.2}$$

$$y(k+1) = f(x(k), u(k)) - x(k) + y(k) \qquad y(0) = y(N) = 0 \tag{5.3}$$

$$z(k+1) = g(x(k), u(k)) + z(k) \qquad z(0) = 0 . \tag{5.4}$$

Thus, the periodic optimization problem has been led to a standard formulation with initial manifold

$$S_0 = \left\{ s : s = | \sigma' \; 0'_{n+1} | , \sigma \in R^n \right\} \tag{5.5}$$

and final manifold

$$S_N = \left\{ s : s = | \sigma' \; 0'_n \; \varrho |' , (\sigma, \varrho) \in R^n \times R \right\} \tag{5.6}$$

where, for all q , 0_q denotes the zero q-vector .
For problem (5.1)-(5.6) let

$$\hat{H}(x(k), y(k), z(k), \alpha(k+1), \beta(k+1), \gamma(k+1), u(k)) = \alpha'(k+1) f(k, u(k)) + \\ + \beta'(k+1)\left[f(x(k), u(k)) - x(k) + y(k)\right] + \gamma(k+1)\left[g(x(k), u(k)) + z(k)\right] \tag{5.7}$$

and assume that $u^\circ(\cdot), x^\circ(\cdot), y^\circ(\cdot), z^\circ(\cdot)$ is an optimal solution.

Then in view of [8, Theorem 2] there must exist $\alpha^\circ(\cdot)$, $\beta^\circ(\cdot)$, $\gamma^\circ(\cdot)$ such that

$$\delta^1 \hat{H}(x^\circ(k), y^\circ(k), z^\circ(k), \alpha^\circ(k+1), \beta^\circ(k+1), \gamma^\circ(k+1), u^\circ(k)) \leq 0 \tag{5.8}$$

$$\alpha^{\circ}(k) = f'_x(x^{\circ}(k),u^{\circ}(k))\alpha^{\circ}(k+1) + \left[f'_x(x^{\circ}(k),u^{\circ}(k)) - I\right]\beta^{\circ}(k+1) +$$

(5.9) $$+g'_x(x^{\circ}(k),u^{\circ}(k))\gamma^{\circ}(k+1)$$

(5.10) $$\beta^{\circ}(k) = \beta^{\circ}(k+1)$$

(5.11) $$\gamma^{\circ}(k) = \gamma^{\circ}(k+1)$$

while the transversality conditions

(5.12) $$\alpha^{\circ}(0) = \alpha^{\circ}(N) = 0$$

(5.13) $$\gamma^{\circ}(N) = 1$$

are met with.

So far, letting

$$\lambda(k) = \alpha(k) + \beta(k)$$

it easily turns out from (5.2) that

$$\hat{H}(x(k),y(k),z(k),\alpha(k+1),\beta(k+1),\gamma(k+1),u(k) = H(x(k),\lambda(k+1),u(k)) -$$

$$-\beta'(k+1)\left[x(k)-y(k)\right] + \gamma(k+1)\,z(k)$$

and in view of (5.8)-(5.13) the conclusion of the first part of the theorem follows.

Then assume $u^{\circ}(k)$ is an internal point of U and, for any $c \in R^m$, consider the admissible N-periodic weak control variation $\delta\tilde{u}(\cdot)$ defined by

$$\delta\tilde{u}(h) = \varepsilon\,\delta_k^{((h))_N}\, c \tag{5.14}$$

where δ_i^j is the Kronecker delta and $((h))_N$ means h modulo N. Correspondingly, in view of the first part of the theorem, it is easy to see that

$$\delta^1 \mathcal{J} = 0 .$$

Hence, a local necessary optimality condition is given by

$$\delta^2 \mathcal{J} = \frac{1}{2N} \sum_{h=k+1}^{k+N} \left| \delta x'(h) \quad \delta u'(h) \right| \left| \begin{matrix} P(h) & Q(h) \\ Q'(h) & R(h) \end{matrix} \right| \left| \begin{matrix} \delta x(h) \\ \delta u(h) \end{matrix} \right| \leq 0 \tag{5.15}$$

where

$$\delta x(h+1) = A(h)\delta x(h) + B(h)\delta u(h) \tag{5.16}$$

$$\delta x(N) = \delta x(0) .$$

The explicit solution to the N-periodic linear system (5.16) is supplied by

$$\delta x(h) = \Phi(h,j)\,\delta x(j) + \sum_{m=j}^{h-1} \Phi(h,m+1)\, B(m)\, \delta u(m) .$$

Since, by assumption, $I - \Phi(k+N+1,k+1)$ is nonsingular, it is easy to check that the perturbed N-periodic motion $\delta\tilde{x}(\cdot)$ corresponding to $\delta\tilde{u}(\cdot)$ turns out to be

$$\delta\tilde{x}(h) = \varepsilon\, \Phi(h,k+1)\,\Gamma(k,N)\,c, \qquad k+1 \leq h \leq k+N \tag{5.17}$$

in view of (5.14) (5.17), eq. (5.15) becomes

$$\mathcal{J} = \frac{\varepsilon^2}{2N} c' \Big\{ R(k) + \Gamma'(k,N)\, \Phi'(k+N,k+1)\, Q(k) + Q'(k)\, \Phi(k+N,k+1)\, \Gamma(k,N) + \Gamma'(k,N) \cdot \left[\sum_{h=k+1}^{k+N} \Phi'(h,k+1)\, P(h)\, \Phi(h,k+1) \right] \Gamma(k,N) \Big\} c \le 0$$

and it is easy to see that, by letting

$$\Phi'(k+N,k+1)\, Q(k) = \tilde{Q}(k,N)$$

$$\sum_{h=k+1}^{k+N} \Phi'(h,k+1)\, P(h)\, \Phi(h,k+1) = \tilde{P}(k+1,k+N)$$

the conclusions of the second part of the theorem follows.

Necessary optimality conditions in the large will now be obtained by extending [8, Theorem 3] to the periodic case.

Theorem 2 (Maximum Principle)

If

$$R(a) = \left\{ \begin{Vmatrix} x \\ x_0 \end{Vmatrix} : \begin{array}{l} x = f(a,u) \\ x_0 \le g(a,u) \end{array} \quad u \in U \right\}$$

is a convex subset of R^{n+1} for all $a \in R^n$ and if $(u^o(\cdot), x^o(\cdot))$ is an optimal solution, then there must exist $\lambda^o(\cdot)$

such that

$$H(x^0(h), \lambda^0(h+1), u) \leq H(x^0(h), \lambda^0(h+1), u^0(h)) \qquad h=0,1,\dots,N-1$$

for all $u \in U$

and

$$\lambda^0(h) = H'_x(x^0(h), \lambda^0(h+1), u^0(h))$$

$$\lambda^0(N) = \lambda^0(0)$$

Proof

The proof follows the same line of reasoning as in the first part of Theorem 1; i.e. problem (1.1)-(1.4) is reformulated and subsequently Propoi's result applied. The only point which is worthy to be remarked is that for the modified problem the convexity of $R(a)$ for all $a \in R^n$ implies the convexity of the corresponding set

$$\hat{R}(a,b) = \left\{ \begin{vmatrix} x \\ y \\ x_0 \end{vmatrix} : \begin{array}{l} x = f(a,u) \\ y = f(a,u) - a + b, \\ x_0 \leq g(a,u) \end{array} \quad u \in U \right\}$$

for all $(a,b) \in R^n \times R^n$.

6. Concluding Remarks

A discrete periodic optimization problem has been discussed by means of three approaches i.e. Dynamic Prog ramming, Mathematical Programming and mainly the Maximum Principle.

It is worthwhile noticing that nearly all results concern the fixed period problem. In particular no condition at all has been given to the free period problem through the Maximum Principle. This is due to the basic reason that no condition is available in the Discrete Maximum Principle literature on unspecific final time problems. As for the Mathematical Programming approach, particular free period problems have been discussed in [16] under the additional constraint that a given set must be met with by the closed trajectory.

REFERENCES

[1] R. Aris, G.L. Nemhauser, D.J. Wilde, "Optimization of Multistage Cyclic and Branching Systems by Serial Procedures", ANN. Hist. Chem. Engs. J., Vol. 10, N°6, pp. 913-919, Nov. 1964.

[2] U. Bertelé, F. Brioschi: 'Non-serial Dynamic Programming", Academic Press, New-York 1972 (to appear).

[3] D. Tabak: 'Application of Mathematical Programming Techniques in Optimal Control: a Survey" , IEEE Trans. On Autom Contr., Vol. AC-15, N° 6, pp. 688-690, Dec. 1970.

[4] L.S. Pontryagin, V.G. Boltyanskii , R.V. Gamkrelidze, E.F. Mischenko: "The Mathematical Theory of Optimal Processes" (Engl. Transl.) (L.W. Neustadt ed.), Wiley (Interscience), New-York 1962.

[5] S. Katz: "A Discrete Version of Pontryagin's Maximum Principle", J. Electr. Contr. Vol. 13, N° 2, p. 179, 1962.

[6] L. Fan, C. Wang, " The Discrete Maximum Principle", John Wiley and Sons, New-York, 1964.

[7] A.G. Butkowskii, "The Necessary and Sufficient Conditions for Optimality of Discrete Control Systems", Avtomatika i Telemekhanika, 24, N°8 (1963).

[8] A.I. Propoi, " The Maximum Principle for Discrete Control Systems" , Avtomatika i Telemekhanika, Vol 26, N°7, (1965).

[9] A.J.M. Holtzmann, H. Halkin, " Directional Convexity and the Maximum Principle for Discrete Systems",

SIAM, J. Control, Vol. 4, N°2, May 1966.

[10] J.M. Holtzmann, "On the Maximum Principle for Non-linear Discrete-Time Systems", IEEE Trans. on Autom-Contr., Vol. AC-11,N°2, April 1966.

[11] M. Canon, C. Cullum, E. Polak: "Constrained Minimization Problems in Finite-Dimensional Spaces", SIAM J. Contr. Vol. 4, N°3, Aug. 1966.

[12] R. Gabasov, F.M.Kirillova, "Extensions of L.S. Pontriagin Maximum Principle to Discrete Systems" Automat. i Telemekhon. Vol 27, N° 11 (1966).

[13] R. Gabasov, "Theory of Optimal Processes in Discrete Systems", Zh. Vych. Matem i Matemat. Fiz., Vol. 8, N°4 (1968).

[14] R. Gabosov, F.M.Kirillova, "Theory of Necessary Conditions of Optimality for Discrete Systems", Avtom i Telemekhan. Vol. 30, N°12, (1968).

[15] R. Gabasov, N.V. Tarashenko, " Necessary High-Order Conditions of Optimality for Discrete Systems", Avtom i Telemekhan, Vol 32, N° 7 (1971).

[16] C. Maffezzoni, S. Rinaldi, "Periodic Optimization of Discrete Systems" , Ricerche di Automatica Vol. 2, N° 3, December 1971.

SERGIO BITTANTI

GIORGIO FRONZA

GUIDO GUARDABASSI (*)

OPTIMAL STEADY – STATE Vs PERIODIC OPERATION IN DISCRETE SYSTEMS (**)

(*) Istituto di Elettrotecnica ed Elettronica, Politecnico di Milano, Milano, Italy

(**) Supported by Centro di Teoria dei Sistemi (CNR)

1. Problem Statement

In dealing with periodic optimization problems, a somewhat preliminary and important question which has first to be answered is whether standard (optimal) steady-state control can be improved or not by means of a suitable cyclic operation. This question, which has been extensively considered in a lot of application-oriented works (mainly in the field of Chemical Engineering [1] - [5]) does also present a number of attractive features, from a theoretical point of view [6] - [9]. Aim of this lecture is to survey all theoretical achieveme ments of the kind above in the area of Discrete Periodic Optimization, and contemporarily to illustrate them, by means of an elementary example.

To be specific, let us be given a system described by

$$x(k+1) = f(x(k), u(k)) \qquad k = 0,1,\ldots,N-1 \tag{1.1a}$$

$$u(k) \in U \subset R^m \tag{1.1b}$$

$$x(k) \in X = R^n \tag{1.1c}$$

where $f(\cdot,\cdot)$ is a known function from $X \times U$ into X, and consider the problem of maximizing

(1.2) $$J\left[u(\cdot),N\right] \triangleq \frac{1}{N}\sum_{k=0}^{N-1} g\left(x(k),u(k)\right),$$

where $g(\cdot,\cdot)$ is a given function from $X \times U$ into R^1, under the periodicity constraint

(1.3) $$x(N)=x(0).$$

The set

$$\Omega \triangleq \left\{u(\cdot) : u(k)\in U , \qquad k=0,1,\dots,N\right\}$$

will now onwards be referred to as the class of admissible controls.

As it is well known, the optimal constant control of system (1.1) with respect to J can be obtained by solving the following mathematical programming problem:

(1.4) $$\text{maximize} \quad g(x,u)$$

subject to

(1.5) $$x-f(x,u)=0$$

(1.6) $$u\in U\subset R^m.$$

In what follows it is assumed, without any loss of generality, that a solution to problem (1.4)-(1.6) exists and is unique (if no solution exists, the problem here makes no sense, on the other hand the treatment below obviously applies to each possible solution, in the multiple solution case); this solution is denoted by $(\bar{x}^{o}, \bar{u}^{o})$ and referred to as optimal steady state.

Definition 1

The optimal periodic control problem (1.1)-(1.3) is said to be proper if there exist a finite integer $\hat{N}$ and an admissible control $\hat{u}(\cdot)$ such that

$$J\left[\hat{u}(\cdot), \hat{N}\right] > \bar{J}^{o}$$

where $\bar{J}^{o} = g(\bar{x}^{o}, \bar{u}^{o})$ is the value of the performance index at the optimal steady state.

The present chapter will then be mainly concerned with the following

Problem

Find proper periodicity conditions for problem (1.1)-(1.3); i.e. find conditions under which problem (1.1)-(1.3) can be said to be proper.

Basically, this problem may be approached in two substantially different ways:

i) by determining a (possibly suboptimal) solution of problem (1.1)-(1.3) and then comparing it with the optimal steady-state solution,

ii) by a local analysis at the optimal steady-state.

Since the first approach requires a solution of the periodic optimization problem, it apparently takes over the particular question under consideration and will not be further discussed here. As for the second approach, it may generally be said to consist in finding conditions under which the optimal constant control can be proved not to be an optimal one, within the class Ω of admissible controls.

The present exposition is organized as follows. The possibility of exploiting the available discrete versions of the Maximum Principle in order to abtain proper periodicity conditions is first discussed. Specifically, it will be shown that in contrast with the continuous case, the only effective result in this direction can be obtained by means of the Weak Pseudo-Maximum Principle for Discrete Periodic Optimization given in [10] . Then, a frequency-domain condition for the proper periodicity of problem (1.1)-(1.3) is presented and its connections with a number of fundamental problem of Discrete System Theory are briefly pointed out.

2. The Maximum Principle Approach to the Proper Periodicity Problem

In order to properly state the theorems below, it is useful to introduce some preliminary definitions and notations.

The Hamiltonian function $H(\cdot,\cdot,\cdot): X \times R^n \times U \rightarrow R^1$ is defined by

$$H(x, \lambda, u) \triangleq g(x, u) + \lambda' f(x, u). \tag{2.1}$$

Furthermore let

$$A \triangleq f_x(\bar{x}^0, \bar{u}^0), \qquad B \triangleq f_u(\bar{x}^0, \bar{u}^0)$$

(subscripts denote partial derivatives) and assume that the absolute value of no eigenvalue of A is equal to one.

Then, define

$$\bar{\lambda}^0 \triangleq (I - A')^{-1} g'_x(\bar{x}^0, \bar{u}^0) \tag{2.2}$$

where prime denotes transpose. A first and well-known property of the optimal steady state solution of problem (1.1)-(1.3) can now be recalled.

Lemma 1.

To each admissible (weak) control variation around the optimal steady-state, there corresponds a non positive first variation of $H(\bar{x}^0, \bar{\lambda}^0, \cdot)$, i.e.:

$$\delta^1 H(\bar{x}^0, \bar{\lambda}^0, \bar{u}^0) = 0 .$$

Proof.

Immediate from the Kuhn-Tucker theorem, as it applies to problem (1.4)-(1.6).

In view of Lemma 1, the only possibility that a proper periodicity condition be obtained by means of the Weak Pseudo-Maximum Principle must rely on the second order necessary condition. In fact, let us introduce the following notation:

$$P \triangleq H_{xx}(\bar{x}^0, \bar{\lambda}^0, \bar{u}^0), \quad Q \triangleq H_{x,u}(\bar{x}^0, \bar{\lambda}^0, \bar{u}^0), \quad R = H_{uu}(\bar{x}^0, \bar{\lambda}^0, \bar{u}^0)$$

Definition 2

Any $(k \times k)$ - Hermitian matrix M is said to be partially positive if there exists $\sigma \in R^k$ such that $\sigma' M \sigma > 0$

Of course, M is partially positive if and only if it is not negative definite nor negative semidefinite. Then, the following theorem, whose proof is omitted as a trivial exercise of matrix book-keeping, can easily be drawn from the Weak Pseudo-Maximum Principle given in [10] .

Theorem 1

If $\bar{u}^0$ is an internal point of U and the $(m \times m)$ matrix

$$\tilde{R}(N) = R + \Gamma'(N)\tilde{Q}(N) + \tilde{Q}'(N)\Gamma(N) + \Gamma'(N)\tilde{P}(N)\Gamma(N)$$

where

$$\Gamma(N) \triangleq (I - A^N)^{-1} B$$

$$\tilde{P}(N) \triangleq \sum_{k=0}^{N-1} (A')^k P A^k, \qquad \tilde{Q}(N) \triangleq A^{N-1} Q$$

is partially positive for some $N \geq 2$, then the optimal periodic control problem (1.1)-(1.3) is proper.

Remark 1

As for the computation of $\tilde{P}(N)$, the following recurrence relation may also turn out to be useful

$$\tilde{P}(k+1) = A'\,\tilde{P}(k)\,A + P \;, \qquad \tilde{P}(1) = P\,.$$

As an illustration of Theorem 1, let us discuss a very simple example.

<u>Example</u>

Let $m = 1$, $n = 2$, and

$$f\,(x(k), u(k)) = A\,x(k) + b\,u(k)$$

where

$$A = \begin{vmatrix} 0 & 1 \\ 0{,}3 & 0{,}1 \end{vmatrix} \;, \qquad b = \begin{vmatrix} 1 \\ 1{,}2 \end{vmatrix}.$$

Furthermore, let $U = [0,2]$ and

$$g\,(x(k), u(k)) = 3\,u^2(k) + 2u(k) - x_1^2(k)$$

Then, the solution $\bar{u}^0 = 1$, $\bar{x}_1^0 = \bar{x}_2^0 = 2$ for problem (1.4)-(1.6) can easily be found (Fig.1) and, correspondingly

$$P = \begin{vmatrix} -2 & 0 \\ 0 & 0 \end{vmatrix} \;, \quad Q = 0 \;, \quad R = 6\,.$$

Although $H(\bar{x}^0, \bar{\lambda}^0, \cdot)$ attains a minimum at $\bar{u}^0 = 1$ (see Fig. 2), i.e. despite of the positiveness of R , no conclusion can yet be drawn about the proper periodicity of out problem. However, trivial but tedious computations yield:

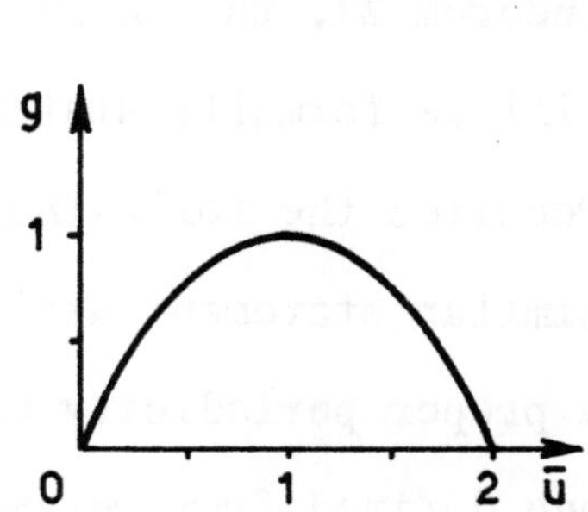

Fig. 1

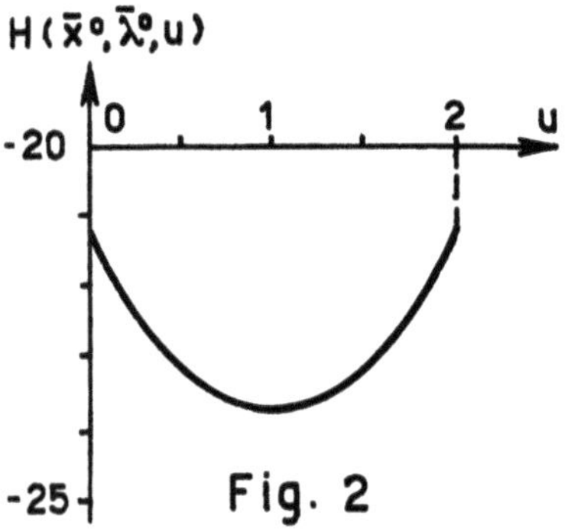

Fig. 2

$$\tilde{R}(2) = -0,25$$

$$\tilde{R}(1) = 1,73$$

Hence, in view of Theorem 1, the conclusion can now be drown that the weak 3-periodic control variation

$$\delta u(k) = \varepsilon \delta_0^{((k))_3},$$

where δ_i^h is the Kronecker delta and $((k))_3$ means k modulo 3, does improve the optimal steady state control $\bar{u}^0 = 1$. Note that a similar conclusion cannot be established for a 2-periodic control variation of the kind above.

So far, a question naturally arises: can Theorem 1 be somehow strenghtened by making use of the strong Maximum Principle for Discrete Periodic Optimization?

Unfortunately, due to the additional convexity requirements concerning functions $f(\cdot,\cdot)$ and $g(\cdot,\cdot)$ ([10], Theorem 2), the answer turns out to be negative. This results will be formally stated as Theorem 2. The proof of Theorem 2 requires the following preliminary lemma, which parallels a similar statement used in [7] to compare, in the continuous case, a proper periodicity based on the Maximum Principle and another one derived from relaxed control theory.

<u>Lemma 2</u>

Lemma 2

For any $u \in U$, consider the function $\gamma(\cdot,\cdot)$: $D \times X \to R^1$, $D = [0,1] \subset R^1$

$$\gamma(\alpha, x) = (1-\alpha)\, g(x, \bar{u}^o) + \alpha\, g(x, u)$$

subject to the equality constraint

$$\varphi(\alpha, x) = 0 \tag{2.3}$$

where the function $\varphi(\cdot,\cdot) : D \times X \to R^n$ is defined as

$$\varphi(\alpha, x) = x + (1-\alpha)\, f(x, \bar{u}^o) + \alpha\, f(x, u). \tag{2.4}$$

Then,

$$\left.\frac{d\gamma}{d\alpha}\right|_{\alpha=0} = H(\bar{x}^o, \bar{\lambda}^o, u) - H(\bar{x}^o, \bar{\lambda}^o, \bar{u}^o) \tag{2.5}$$

where $H(\cdot,\cdot,\cdot)$ and $\bar{\lambda}^o$ are given by eqs. (2.1) and (2.2).

<u>Proof</u>

Note that

$$\left.\frac{d\gamma}{d\alpha}\right|_{\alpha=0} = \left[\frac{\partial\gamma}{\partial\alpha} + \frac{\partial\gamma}{\partial x}\frac{\partial x}{\partial\alpha}\right]_{\alpha=0}. \tag{2.6}$$

On the other hand

$$\left.\frac{\partial\gamma}{\partial\alpha}\right|_{\alpha=0} = g(\bar{x}^o, u) - g(\bar{x}^o, \bar{u}^o) \tag{2.7}$$

$$\left.\frac{\partial\gamma}{\partial x}\right|_{\alpha=0} = g_x(\bar{x}^o, \bar{u}^o) \tag{2.8}$$

while, in view of eq. (2.3) and the assumption on A ,

(2.9) $$\left.\frac{dx}{d\alpha}\right|_{\alpha=0} = (I-A)^{-1}\left(f(\bar{x}^0,u)-f(\bar{x}^0,\bar{u}^0)\right).$$

Thus, recalling eqs. (2.1), (2.2) and substituting (2.7)-(2.9) into (2.6), eq. (2.5) follows.

Theorem 2

The Discrete Maximum Principle for Periodic Optimization cannot be violated at the optimal steady-state.

Proof

By contradiction , if there exists $u \in U$ such that

$$H(\bar{x}^0,\bar{\lambda}^0,u) - H(\bar{x}^0,\bar{\lambda}^0,\bar{u}^0) > 0$$

then, in view of Lemma 2, there exists $(\tilde{\alpha},\tilde{x}) \in D \times X$ such that

(2.10) $$\varphi(\tilde{\alpha},\tilde{x}) = 0$$

and

(2.11) $$\gamma(\tilde{\alpha},\tilde{x}) > \gamma(0,\bar{x}^0) = g(\bar{x}^0,\bar{u}^0).$$

But

$$\varphi(\tilde{\alpha},\tilde{x}) = \tilde{x} + (1-\tilde{\alpha})f(\tilde{x},\bar{u}^0) + \tilde{\alpha}f(\tilde{x},u) =$$

$$= (1-\tilde{\alpha})\left(\tilde{x} - f(\tilde{x},\bar{u}^{o})\right) + \tilde{\alpha}\left(\tilde{x} - f(\tilde{x},u)\right).$$

Hence, (2.10) and (2.11) imply (see fig. 3) that

$$R_1(\tilde{x}) \triangleq \left\{ r : r = \begin{vmatrix} \sigma \\ \varrho \end{vmatrix}, \sigma = f(\tilde{x},u), \ \varrho \le g(\tilde{x},u), u \in U \right\}$$

is not convex, i.e. the basic assumption for the strong Discrete Maximum Principle cannot be verified.

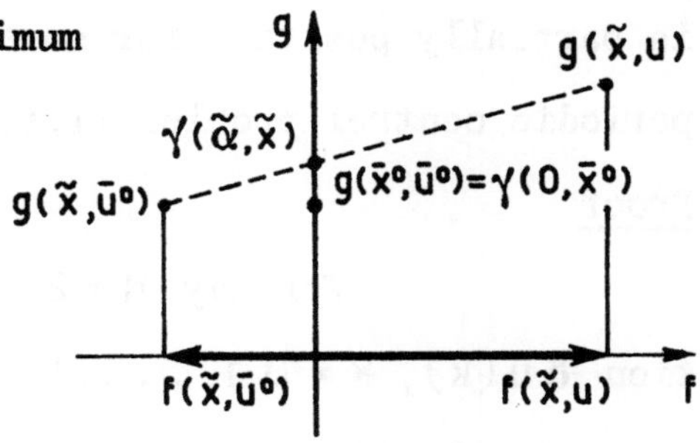

Fig. 3

3. A Frequency Domain Proper Periodicity Condition

Aim of this section is to obtain, through a second variation analysis, a proper periodicity condition in the frequency domain which is shown to encompass the preceding one derived via the Weak Pseudo-Maximum Principle and, in the particular case of single-input single-output systems, can be given the useful form of a circle criterion. In general, it can be stated as follows.

Theorem 3

If $\bar{u}^0$ is an internal point of U and the matrix

$$\Pi(\vartheta) \triangleq R + G'(e^{-j\vartheta})Q + Q'G(e^{j\vartheta}) + G'(e^{-j\vartheta})PG(e^{j\vartheta}), \tag{3.1}$$

where

$$G(z) \triangleq (zI - A)^{-1}B,$$

is partially positive for some $\vartheta \in (0, \pi]$, then the optimal periodic control problem (1.1)-(1.3) is proper.

<u>Proof</u>

For any $N \geq 2$, consider a weak control variation $\delta u(k)$, $k = 0, 1, \ldots, N-1$. Correspondingly, in view of Lemma 1, the first variation of the performance index (1.2) around the optimal steady-state is zero, the second variation is given by

$$\delta^2 J = \frac{1}{2N} \sum_{k=0}^{N-1} \begin{vmatrix} \delta x'(k) & \delta u'(k) \end{vmatrix} \begin{vmatrix} P & Q \\ Q' & R \end{vmatrix} \begin{vmatrix} \delta x(k) \\ \delta u(k) \end{vmatrix} \tag{3.2}$$

where

$$\delta x(k+1) = A\delta x(k) + B\delta u(k), \quad k = 0, 1, \ldots, N-1, \tag{3.3}$$

$$\delta x(N) = \delta x(0) \tag{3.4}$$

Let $w(\cdot)$ be the Discrete Fourier Transform of $\delta u(\cdot)$, i.e.

$$w(h) = \sum_{k=0}^{N-1} \delta u(k) e^{-jk\varphi h}, \quad \varphi = \frac{2\pi}{N}, \quad h = 0, 1, \ldots, N-1$$

so that

$$\delta u(k) = \frac{1}{N} \sum_{h=0}^{N-1} w(h) e^{jh\varphi k}. \tag{3.5}$$

Since it has been assumed that the absolute value of no eigenvalue of A is equal to one, the N -periodic (eq. (3.4)) solution of system (3.3) exists and can be written as

$$\delta x(k) = \frac{1}{N} \sum_{h=0}^{N-1} \xi(h) e^{jh\varphi k}, \tag{3.6}$$

when the Discrete Fourier Transform $\xi(\cdot)$ of $\delta x(\cdot)$ is given by

$$\xi(h) = G(e^{jh\varphi}) w(h). \tag{3.7}$$

Eq. (3.2) can obviously be developed as

$$\delta^2 J = \frac{1}{2N} \sum_{k=0}^{N-1} \delta x'(k) P \delta x(k) + \sum_{k=0}^{N-1} \delta x'(k) Q \delta u(k) +$$

$$+ \sum_{k=0}^{N-1} \delta u'(k) Q' \delta x(k) + \sum_{k=0}^{N-1} \delta u'(k) R \delta u(k)$$

and, by substitution of eq. (3.6), the first term becomes:

$$\sum_{k=0}^{N-1} \delta x'(k) P \delta x(k) = \sum_{k=0}^{N-1} \left[\frac{1}{N} \sum_{r=0}^{N-1} \xi(r) e^{jr\varphi k}\right]' P \left[\frac{1}{N} \sum_{s=0}^{N-1} \xi(s) e^{js\varphi k}\right]$$

$$= \frac{1}{N} \sum_{r=0}^{N-1} \sum_{s=0}^{N-1} \xi'(r) P \xi(s) \frac{1}{N} \sum_{k=0}^{N-1} e^{j(r+s)\varphi k}.$$

But

$$\sum_{k=0}^{N-1} e^{jh\varphi k} = N\delta_0^{((h))_N},$$

where δ_i^k is the Kronecker delta and $((h))_N$ means h modulo N, hence

$$\sum_{k=0}^{N-1} \delta x'(k) P \,\delta x(k) = \frac{1}{N}\Big[\xi'(0)P\,\xi(0) + \sum_{h=0}^{N-1} \xi'(N-h)P\xi(h)\Big] =$$

$$= \frac{1}{N}\sum_{k=0}^{N-1} \xi^*(h)P\xi(h)$$

where star denotes conjugate transpose. Finally, in view of eq. (3.7)

$$\sum_{k=0}^{N-1} \delta x'(k) P\,\delta x(k) = \frac{1}{N}\sum_{h=0}^{N-1} w^*(h)G'(e^{-jh\varphi})\,PG(e^{jh\varphi})w(h).$$

By following quite similar arguments, it may be concluded that

$$\delta^2 J = \frac{1}{2N^2}\sum_{h=0}^{N-1} w^*(h)\,\Pi\,(h\varphi)w(h) \tag{3.8}$$

where $\Pi(\cdot)$ is given by (3.1). By continuity, if $\Pi(\vartheta)$ is partially positive for some $\vartheta \in (0,\pi]$, then there esist $\hat{h}$ and $\hat{N}$ such that $\Pi(\hat{h}\,\hat{\varphi})$, where $\hat{\varphi} \triangleq \frac{2\pi}{\hat{N}}$, is partially positive too. Let $\chi \in R^m$ be such that

$$\chi'\,\Pi\,(\hat{h}\,\hat{\varphi})\chi > 0,$$

then the $\hat{N}$ -periodic weak control variation corresponding to

$$\hat{w}(h) = \frac{1}{2}\chi\,\delta^{h}_{\hat{h}}\ , \qquad h = 0,1,\ldots,N-1$$

i.e. (see eq. 3.3))

$$\widehat{\delta u}(k) = \varepsilon\chi\cos\hat{h}\hat{\varphi}k$$

apparently improves the performance index at the optimal steady-state.

So far, we have shown that $\delta^2 J > 0$ is equivalent to the partial positivity of $\Pi(\vartheta)$ for some $\vartheta \in (0,\pi]$; on the other hand, in view of Theorem 1, the positivity of the second variation is only implied by the partial positivity of $\tilde{R}(N)$ for some $N \geq 2$ (weaker condition). A deeper insight into the existing relationship between Theorem 1 and Theorem 3 is supplied by the following results.

<u>Remark 2</u>

When N is fixed, Theorem 3 can obviously be rephrased as follows. If $\bar{u}^0$ is an internal point of U and $\Pi\left(\frac{2\pi}{N}h\right)$ is partially positive for some $h = 1,2,\ldots,N-1$ (note that since $\bar{u}^0$ is the optimal constant control then $\Pi(0) \leq 0$) , then there exists a weak N-periodic control variation $\widehat{\delta u}(\cdot)$ which improves the optimal constant control.

<u>Remark 3</u>

Letting $\tilde{R}(\cdot)$ and $\Pi(\cdot)$ be defined as in Theorem 1 and Theorem 3, respectively, then

$$\tilde{R}(N) = \frac{1}{N} \sum_{h=0}^{N-1} \Pi\left(\frac{2\pi}{N} h\right).$$

Proof

For any $w \in R^m$, consider a control variation

$$\delta\hat{u}(k) = w\, \delta_0^{((k))_N}.$$

The corresponding second variation of the performance index at the optimal steady state is given by (see [10] , Theorem 1)

$$\delta^2 J = \frac{1}{2N} w' \tilde{R}(N) w . \tag{3.9}$$

On the other hand, the Discrete Fourier Transform of $\delta\hat{u}(\cdot)$ is a constant, i.e.

$$w(h) = w , \quad h = 0,1,\dots,N-1 ;$$

hence, in view of eq. (3. 8),

$$\delta^2 J = \frac{1}{2N^2} w' \left(\sum_{h=0}^{N-1} \left(\frac{2\pi}{N} h\right)\right) w . \tag{3.10}$$

Since eqs. (3.8) and (3.10) hold for any $w \in R^m$, the theorem follows.

Example

Remark 3 immediately shows that if $\tilde{R}(N)$ is partially positive for some $N \geq 2$, then $\Pi\left(\frac{2\pi}{N} h\right)$ must be partially positive too, for some $h = 1, 2, \dots, N-1$. However, it should also be apparent that in general the converse is not true. As a matter of fact, let us consider again the elementary example used in the preceding section to illustrate Theorem 1. Quite standard computations yield

$$\Pi(\vartheta) = 6 - 2\left|G_1(e^{i\vartheta})\right|^2,$$

where

$$G_1(z) = \frac{1{,}2}{z^2 - 0{,}1z - 0{,}3},$$

and it is easy to check that $\Pi(\pi) = 1{,}5 > 0$; hence, in view of Remark 2 and Theorem 3, the optimal steady-state operation can apparently be improved by means of a weak 2-periodic control variation of the kind

$$\delta u(k) = \varepsilon \cos k\pi = \varepsilon(-1)^k.$$

It is worth recalling that such a result could not be achieved by making use of Theorem 1 $(\tilde{R}(2) < 0)$.

Finally, let us restrict to a special class of systems, namely single-input single-output linear systems; then, the general condition stated by Theorems 3 can be given a single and useful geometrical interpretation.

Consider the system

$$x(k+1) = Ax(k) + bu(k) \tag{3.11a}$$

$$y(k) = c'x(k) \tag{3.11b}$$

(3.12) $$u(k) \in U \subset R^1, \quad x(k) \in X = R^n, \quad y(k) \in Y = R^1$$

A, b, c = real constants matrices of suitable order subject to the periodicity constraint

(3.13) $$x(N) = x(0)$$

and assume that the absolute value of no eingenvalue of A is equal to one. Let the performance index J be of the form

(3.14) $$J[u(\cdot), N] = \frac{1}{N} \sum_{k=0}^{N-1} \eta(y(k), u(k))$$

where $\eta(\cdot,\cdot)$ is a given twice differentiable real function. Furthermore let

$$p \triangleq \eta_{yy}(\bar{y}^o, \bar{u}^o), \quad q \triangleq \eta_{yu}(\bar{y}^o, \bar{u}^o), \quad r \triangleq \eta_{uu}(\bar{y}^o, \bar{u}^o)$$

where $\bar{y}^o = c' \bar{x}^o$,and, for any triplet (p , q , r) such that $q^2 - rp \geq 0$, define the critical circle in the complex plane as the circle of radius

$$\varrho = \sqrt{\left(\frac{q}{p}\right)^2 - \frac{r}{p}}$$

and center on the real exis at

$$\sigma = -\frac{q}{p} .$$

When p vanishes, the critical circle degenerates in a straight line, parallel to the imaginary axis and crossing the real axis at $\tilde{\sigma} = -r/2q$.

The critical region is then defined as the open set of all points outside or inside the critical circle, according to whether p is positive or negative $(p \neq 0)$. When $p = 0$, the critical region is the right or left open half plane defined by the vertical "critical line", according to whether q is positive or not. The case $q = p = 0$ is apparently trivial.

Then, the following theorem can be stated.

Theorem 4

Let

$$F(z) = c'(zI - A)^{-1} b$$

be the transfer function of system (3.11). If the polar plot of $F(e^{i\vartheta}), \vartheta \in (0, \pi]$, enters the critical region, then the periodic optimization problem (3.11)-(3.12) is proper.

Proof

In view of the definitions above, it is easy to see that $\Pi(\vartheta)$ is here a scalar which can be put in the form

$$\Pi(\vartheta) = r + 2q \operatorname{Re}\left[F(e^{i\vartheta})\right] + p \left|F(e^{i\vartheta})\right|^2 .$$

Tedious but trivial computations show that the locus in the F-plane where $\Pi(\vartheta) > 0$ is just the critical region. Hence, the theorem is proved.

Example

Let us consider again the elementary example described in Sect.2. Letting $c' = |1 \ 0|$, we have

$$\eta(y(k), u(k)) = 3u^2(k) + 2u(k) - y^2(k)$$

and $\bar{y}^0 = 2$. Furthermore

$$p = -2 \ , \quad q = 0 \ , \quad r = 6$$

$$F(z) = G_1(z) = \frac{1{,}2}{z^2 - 0{,}1z - 0{,}3}$$

so that

$$\varrho = \sqrt{3} \ , \quad \sigma = 0$$

and the application of the circle criterion is illustrated in Fig. 4.

4. Concluding Remarks

Optimal steady-state and proper periodic operations for discrete dynamical systems have been compared in this paper and sufficient conditions for the latter to be better than the former have been pointed out. In particular, the frequency-domain condition derived in Sec. 3 has been proved to be the strongest available criterion, for the dis-

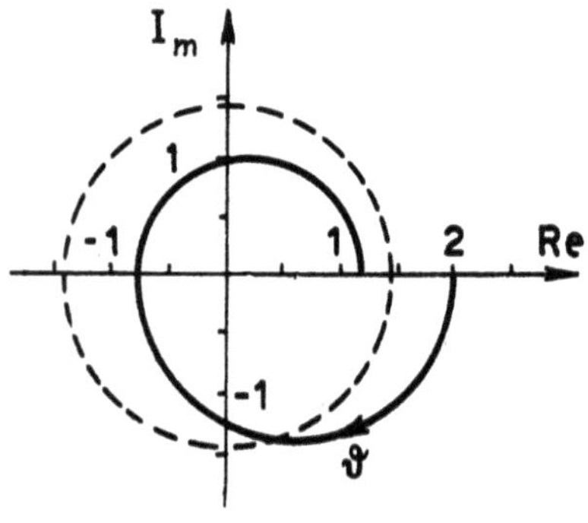

Fig. 4

crete case, although the time-domain condition given in Sect. 2 is generally simpler from a computational point of view. This is not the case, of course, for special classes of problems such as the one considered in Theorem 4, where the frequency domain condition has been given the useful form of a circle criterion.

As in the continuous case, Theorem 3 points out important relationships between a number of basic problems od Feedback System Theory (such as absolute stability [11] , [12] , sensitivity [13] , existence of solutions for algebraic Riccati equations of the kind encountered in linear quadratic optimal control theory [14] , and so on) and the problem considered here. However, the expected non-trivial differences between the discrete and the continuous case, basically due to the possible singularity of the transition matrix in the discrete case (non-reversibility), are not yet sufficiently clarified and constitute an interesting field for further researches.

Another open and interesting problem is the determination of efficient algorithms for testing the partial positivity condition given by Theorem 3. Even in this connection, any possible relation with the unsolvability of algebric linear or non linear (Riccati) matrix equations might play a fundamental role.

Finally, the effectiveness of the result presented in this lecture should be conveniently ascertained in front of some of the many relevant applications reported in the literature. In other words, it is conceivable to guess that proper periodic operations, improving the optimal steady state but often discovered through purely empiric or semi-empiric procedures, should be further substantially improved by making suitable use of all the available information about the harmonic response of the system at the optimal steady-state.

REFERENCES

[1] M.H.I. Baird, "Vibration and Pulsation", British Chem. Eng., Vol. 11, N°1, pp. 20-25, 1966.

[2] Symposium on Cyclic Processing Operations, I and EC Proc. Des. and Development, Vol. 6, N°1, 1967.

[3] J.E. Bailey, F.J.M. Horn, R.C. Lin, " Cyclic Operation of Reaction Systems" , 63rd Nat. AIChE Meeting, St. Louis, Missouri, 1968.

[4] F.J.M. Horn, J.E. Bailey, " An Application of the Theorem of Relaxed Control to the Problem of Improving Catalyst Selecting, JOTA, Vol.2 N°5, 1968.

[5] N.Y. Gaitonde, J.M. Douglas, " The Use of Positive Feed back Control Systems to Improve Reactor Performance" , AIChE Journal, Vol. 15, Vol. 15, N°6, pp. 902-910, 1969.

[6] F.J.M. Horn, R.C. Lin, " Periodic Processes: A Variational Approach" , I and EC Proc. Des. and Development, Vol. 6, N°1, pp. 21-30, 1967.

[7] J.E. Bailey, F.J.M. Horn, " Comparison between Two Sufficient Conditions for Improvement of an Optimal Steady-State Process by Periodic Operation" , JOTA, Vol. 7, N°5, pp. 378-385, 1971.

[8] G. Guardabassi, " Optimal Steady-State Versus Periodic Control: a Circle Criterion", Ricerche di Automatica, Vol. 2, N°3, December 1971.

[9] S. Bittanti, G. Fronza, G. Guardabassi, " Periodic Control: a Frequency Domain Approach" ,

IEEE Trans. on Autom. Contr., Vol. AC-18, N°1, february 1973.

[10] S. Bittanti, G. Fronza, G. Guardabassi, "Optimality Conditions for Discrete Periodic Optimization", this volume.

[11] E.S. Pyatnitskii, "New Research on the Absolute Stability of Automatic Control Systems (Review)", Avtomatika i Telemekhanika, N°6, pp. 5-36, 1968.

[12] Ya.Z. Tsypkin, "Frequency Criteria of Absolute Stability of Nonlinear Pulse Systems", Automatika i Telemekhanika, Vol. 25, N°3, 1964.

[13] J.B. Cruz, W.R. Perkins, "A New Approach to the Sensitivity Problem in Multivariable Feedback System Design", IEEE Trans. on Automatic Control, Vol. AC-9, pp. 216-223, 1964.

[14] J.C. Williams, "Least Square Stationary Optimal Control and the Algebraic Riccati Equation", IEEE Trans. on Automatic Control, Vol. AC-16, pp. 621-634, 1971.

ARTURO LOCATELLI

CLAUDIO MAFFEZZONI

SERGIO RINALDI (*)

A SEARCHING PROCEDURE FOR THE OPTIMAL FREQUENCY IN PERIODIC OPTIMIZATION (**)

(*) Istituto di Elettrotecnica ed Elettronica, Politecnico di Milano, Milano, Italy
(**) Supported by Centro di Teoria dei Sistemi (CNR)

1. Introduction

One of the main problems which has to be solved in Periodic Optimization is the determination of the optimal frequency.

A classical approach relies on necessary and/or sufficient conditions for optimality which are used as a starting point for suitable computing methods. This has been done, for example, by Chang [1] in the case of continuous systems.

However, when optimality conditions for the frequency are not available (as for discrete systems) or an effective computational procedure for fixed-period problems exists, one might determine the optimal frequency by solving a sequence of optimization problems with fixed period. Of course, a good searching procedure must be known in order to keep the computational effort within reasonable limits. In other words, the optimal value of the performance as a function of the period must exhibit suitable structural properties so that a satisfactory seeking method could be applied. For example, if the cost function were unimodal on a certain interval, then the well-known Fibonacci's search might be properly used for the determination of the optimal period.

Unfortunately, no sufficiently nice structural properties (like unimodality) of the cost function have been proved up to now in the general case. However, for a particular class of periodic optimization problems (i.e. those where the optimal cycle is constrained to meet a given equilibrium state) an elimination procedure has been found both for continuous [2] and discrete systems [3]. The idea underneath is the same in both cases, the only relevant difference being the necessity of a proof of convergence when dealing with continuous systems. For this reason only the discrete case is here presented.

2. Statement of the Problem

Consider a time-invariant discrete system described by a set of difference equations of the form

(2.1) $$x(k+1) = f(x(k), u(k)) \qquad k = 0,1,2,\ldots$$

where

(2.2) $$u \in U \subset R^m \qquad \text{(control vector)}$$

(2.3) $$x \in X \subset R^n \qquad \text{(state vector)}.$$

Then, minimize an average performance index, i.e.

(2.4) $$c^\circ = \min \frac{1}{n} \sum_{i=0}^{n-1} g(x(i), u(i)) \ ,$$

subject to the " periodicity constraint"

(2.5) $$x(0) = x(n) = \bar{x} \qquad \text{(given)}$$

where the period n is such that

$$n \in \mathcal{N} = \{1, 2, \ldots, n_M\} . \tag{2.6}$$

With reference to the discrete periodic optimization problem, it must be noted that the optimal cycle of length $n \leq \bar{n}$ can not be obtained (as when applying Dynamic Programming) by solving the mathematical program (2.1)-(2.5), for $n = \bar{n}$. Hence n_M different mathematical programs should be solved. In this connection, an important question consists in asking under what conditions it is possible to save computations in determining the optimal solution, i.e. to solve the mathematical program (2.1)-(2.5) only for some suitable $n \in \mathcal{N}$.

3. A Secondary Optimization Problem

Problem (2.1)-(2.6) is now splitted in a two levels optimization problem as follows.

<u>First level</u>

$$c(n) = \frac{1}{n} \min \sum_{i=0}^{n-1} g(x(i), u(i)) \tag{3.1a}$$

subject to

$$u(k) \in U \qquad k = 0, 1, \ldots, n-1 \tag{3.1b}$$

$$x(k) \in X \qquad k = 0, 1, \ldots, n-1 \tag{3.1c}$$

(3.1d) $$x(k+1) = f(x(k), u(k)) \qquad k = 0, 1, \ldots, n-1$$

(3.1e) $$x(n) = x(0) = \bar{x} \, .$$

Second level

(3.2) $$c^{\circ} = \min_{1 \leq n \leq n_M} c(n).$$

The first level optimization problem is assumed to have a solution for all $n \in \mathcal{N}$; hence, for the existence of a solution for $n=1$ $\bar{x}$ must obviously be an equilibrium state . In many cases of practical interest every equilibrium state $\bar{x}$ corresponds to a unique control vector $\bar{u}$, so that $c(1)$ is simply determined by feasibility arguments. In other cases, $c(1)$ is determined by actually solving the optimization problem (3.1).

In the following, however, $c(1)$ is assumed to be known and when talking about k evaluations of the function $c(\cdot)$ at the points $n^{(1)}$, $n^{(2)}$, ..., $n^{(k)}$ the assumption $n^{(i)} \neq 1$, $i = 1, \ldots, k$ will always be understood.

Aim of this section is to state a secondary optimization problem in a way such that its solution allows to solve problem (3.2) withoun evaluating $c(\cdot)$ by means of (3.1) for all $n \in \mathcal{N}$. To this purpose, assume the first level optimization problem (3.1) has been solved for k different periods $n^{(1)}$, $n^{(2)}$, ..., $n^{(k)}$; then, an upper bound for c° is given by the " minimum at hand" $c_m^{(k)} = \min_{1 \le i \le k} c\left(n^{(i)}\right)$.

Moreover, suppose that a lower bound $\bar{c}^{(k)}$ of c^o, generally dependent upon the first k evaluations of $c(\cdot)$, is also available: obviously, this will always be the case if a lower bound $\Theta > 0$ of c^o is a priori known, as it is assumed in the following (some comments relative to other cases can be found at the end of Sect. 5). Thus, the ratio

$$\eta^{(k)} = \frac{c_m^{(k)}}{\bar{c}^{(k)}}$$

greater than or equal to one, is a significant index of the degree of suboptimality of the minimum at hand. Finally, note that after $c(\cdot)$ has been evaluated for k different values of n, a set $\mathcal{C}_n^{(k)}$ representing the possible values of $c(n)$ (from now on indicated by $c_n^{(k)}$) can be associated to every $n \in \mathcal{N}$.

Thus, the secondary optimization problem consists in determining the integer $n^{(k+1)}$ so that the index $\eta^{(k+1)}$ be minimized in the worst possible case, i.e.

$$\min_{n \in \mathcal{N}} \max_{c_n^{(k)} \in \mathcal{C}_n^{(k)}} \eta^{(k+1)} . \tag{3.3}$$

If criterion (3.3) does not entail a unique solution, in the set $\bar{\mathcal{N}}^{(k)}$ of solutions of (3.3) the optimal one will be determined by the criterion

$$\min_{n \in \bar{\mathcal{N}}^{(k)}} E\left\{\eta^{(k+1)}\right\} \tag{3.4}$$

where the expectation is taken with respect to $c_n^{(k)}$ assuming a uniform probability density function in the set $\mathcal{C}_n^{(k)}$ of admissible variations.

4. Properties of the Cost Function

Some simple properties of the cost function defined by (3.1) are now stated in order to solve the secondary optimization problem.

Property 1

$$(n_1+n_2)\,c\,(n_1+n_2) \leq n_1\,c\,(n_1)+n_2\,c(n_2). \qquad ▨$$

Proof. By concatenation of the two optimal solutions relative to $n=n_1$ and $n=n_2$ an admissible solution for problem (3.1) with $n=n_1+n_2$ is obtained. On the other hand, this solution is, in general, not the optimal one for the last problem. ▨

Property 2

$$c\,(m\;n) \leq c\,(n). \qquad ▨$$

Proof. By applying Property 1 recursively, it turns out to be:

$$m\,n\,c\,(m\;n) = (n+(m-1)n)\,c\,(n+(m-1)n) \leq n\;c\,(n)+$$
$$+(m-1)\,n\;c((m-1)\,n) \leq \;\dots\; \leq m\,n\;c\,(n)$$

and Property 2 follows. ▨

Property 3

$$c^{o} = \min_{\left[\frac{n_M}{2}\right]+1 \leq n \leq n_M} c(n)$$

where $[x]$ = integer part of x. ▨

Proof. Corresponding to every $n<\left[\frac{n_M}{2}\right]+1$ there exists an integer

$m\ (\geq 2)$ such that $\left[\frac{n_M}{2}\right]+1 \leq m\ n \leq n_M$. Thus, from Property 2, Property 3 follows. ▨

<u>Property 4</u>

$$c(n) \leq c(1).$$ ▨

<u>Proof.</u> Trivial from Property 2. ▨

<u>Property 5</u>

$$\frac{c(1)-c(n+1)}{\nu_{n+1}} \geq \frac{c(1)-c(n)}{\nu_n}$$

where $\nu_k = \frac{1}{k}$, with k = integer. ▨

<u>Proof.</u> From Property 1 with $n_1 = 1$ and $n_2 = n$ it follows:

$$(n+1)\,c(n+1) \leq c(1) + n\,c(n).$$

By adding the quantity $n\,c(1)$ to both sides of the above equation, Property 5 is immediately proved. ▨

The last two properties are shown in Fig. 1 where the function $c(\cdot)$ has been evaluated at $\bar{n}$: no point $(\nu_n, c(n))$ falls in the dashed regions. On the other hand, Property 3 allows to reduce the set of possible values of n^0, before making any evaluation of the function $c(\cdot)$.

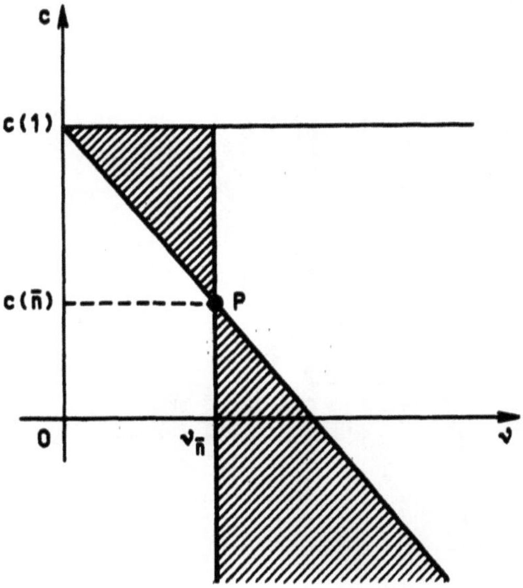

Fig. 1

In fact,

$$n^{0} \in \mathcal{N}^{(0)} \triangleq \left\{ \left[\frac{n_M}{2} \right] + 1, \ldots, n_M \right\} \subset \mathcal{N}.$$

Moreover, after k evaluations of $c(\cdot)$ have been performed, a set $\mathcal{N}^{(k)}$ of feasible values for the optimum period n^0 can be determined together with the value $m^{(k)}$, corresponding to the minimum at hand $c_m^{(k)}$. Fig. 2 shows the fundamental rule for determining $\mathcal{N}^{(k)}$, in the simple case $k=3$. The points P, Q and R represent the three evaluations of $c(\cdot)$, which have been made, of course, at values $n \in \mathcal{N}^{(0)}$. In Fig. 2a the regions where no point $(\nu_n, c(n))$ can fall are dashed. On the other hand, if the function $c(\cdot)$ had been evaluated at any n such that $\nu_{min} < \nu_n \leq \nu'$ or $\nu_h < \nu_n \leq \nu''$ (see dashed regions of Fig. 2b),

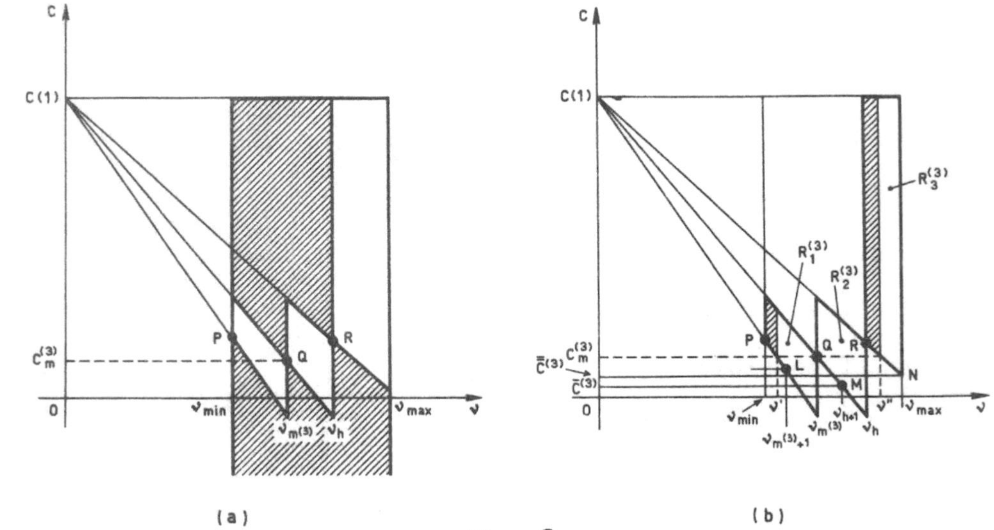

Fig. 2

then values of $c(n)$ lower than $c_m^{(3)}$ would not be obtained.

Thus the point (ν_{n^0}, c^0) lies in one of the three regions $R_1^{(3)}$, $R_2^{(3)}$, $R_3^{(3)}$ of Fig. 2b, which are characterized by the same level $c_m^{(3)}$ of the left low corner and $\mathcal{n}^{(3)}$ is the set of the integers relative to these regions.

This situation is general indeed. That is, when $c(\cdot)$ has been evaluated at k points, the optimal pair (ν_{n^0}, c^0) either coincides with $(\nu_{m(k)}, c_m^{(k)})$ or is contained in the region $R^{(k)}$ consisting of the union of the regions $R_i^{(k)}$, $i = 1, \ldots, N_k$, which can be obtained first by means of Properties 4 and 5 (Fig. 2a) and then by eliminating the subregions lying above the level $c_m^{(k)}$ (Fig. 2b). Moreover, let $\mathcal{n}^{(k)}$ be the union of the sets $\mathcal{n}_i^{(k)}$ which consist of the values ν_n corresponding to the region $R_i^{(k)}$, except the ones where the function $c(\cdot)$ has already been evaluated.

It should be noted that, if one evaluation has been made at ν_{min} (in the following for shortness $\nu_{min} = \frac{1}{n_M}$, $\nu_{max} = \frac{1}{\left[\frac{n_M}{2}\right] + 1}$), the regions $R_i^{(k)}$, $i = 1, \ldots, N_k$, are bounded from below. This ensures the existence of a minimum of $c_n^{(k)}$ for $n \in \mathcal{n}_i^{(k)}$, for every i. Let $\bar{c}_i^{(k)}$ be such a minimum (points L, M and N of Fig. 2) and $\bar{c}^{(k)}$ (see point M of Fig. 2) the minimum among the values $\bar{c}_i^{(k)}$, $i = 1, \ldots, N_k$. The region corresponding to the latter minimum will be denoted by $i^{*(k)}$ ($i^{*(k)} = 2$ in Fig. 2). Moreover, let $\bar{\bar{c}}^{(k)}$ ($\bar{\bar{c}}^{(k)} > \bar{c}^{(k)}$ in the following) be the minimum among the values $\bar{c}_i^{(k)}$ except $\bar{c}^{(k)}$ (see point N of Fig. 2); of course $\bar{\bar{c}}^{(k)}$ exists only if $N_k > 1$.

Finally, in the case where a lower bound Θ of c^{o} is, a priori, known, the regions $R_i^{(k)}$ can be further restricted by eliminating the parts lying below the bound Θ (see Fig. 3). Thus the knowledge of Θ may improve the lower bound $\bar{c}^{(k)}$.

5. Solution of the Secondary Optimization Problem

The solution of the secondary optimization problem is now derived by means of the properties of the cost function discussed in the preceding section. The explicit solution can be given only for some particular cases. Nevertheless, the problem is in general, reduced to the simpler one of determining the minimum of a function defined on a finite number of points.

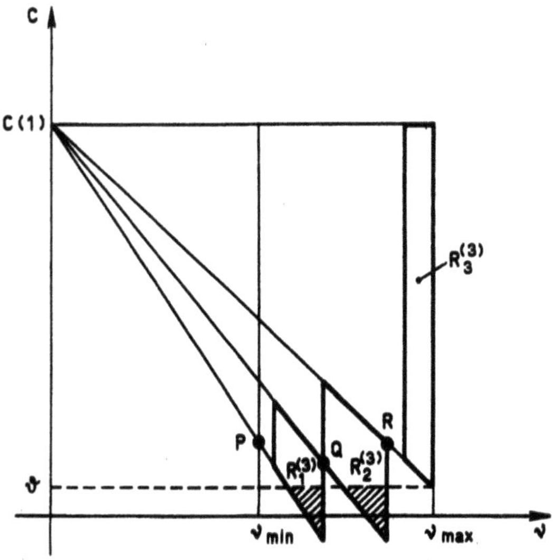

Fig. 3

The flow-chart for solving the secondary optimization problem (i.e. determining $n^{(k+1)}$) is shown in Fig. 4. The starting block A is concerned with the first evaluation of $c(\cdot)$ and states that, in such a case, the solution of the problem is

$$n^{(1)} = n_M . \tag{5.1}$$

Eq. (5.1) is immediate to check since

$$\max_{c_{n_M}^{(0)} \in \mathcal{C}_{n_M}^{(0)}} \eta^{(1)} < \max_{\substack{c_n^{(0)} \in \mathcal{C}_n^{(0)} \\ n < n_M}} \eta^{(1)} = \frac{c(1)}{\Theta} .$$

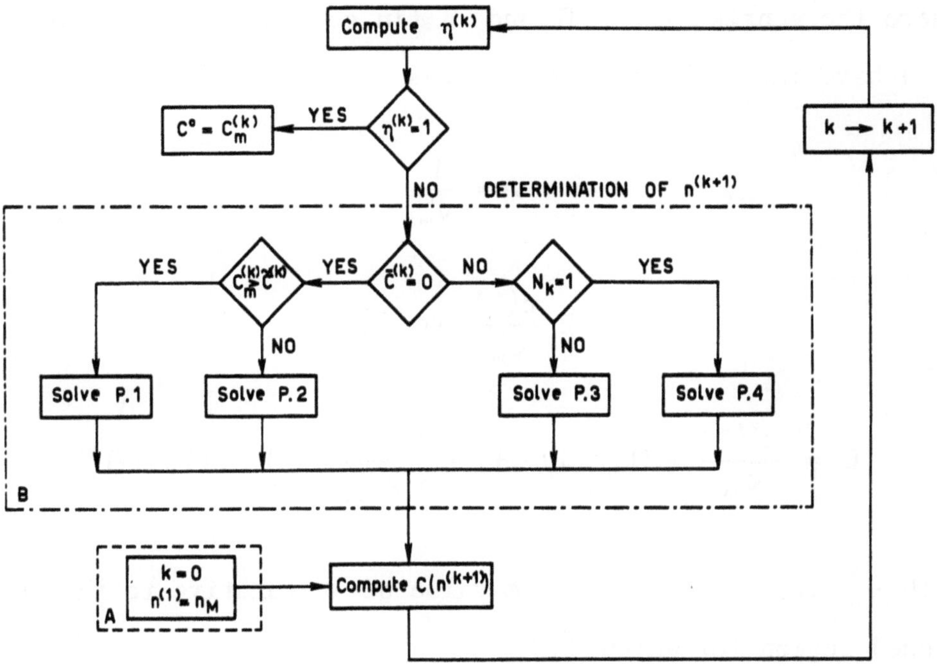

Fig. 4

The block B points out four possible cases for the determination of $n^{(k+1)}$. For every one of these cases a simple minimization must be performed (solution of problems $P_1, \ldots, P_4$). The proof of this fact consists of cumbersome computations and is available in [4].

The quantity $\tilde{c}^{(k)}$, which appears in the block scheme of Fig. 4 is given by

$$\tilde{c}^{(k)} = \frac{B - \sqrt{B^2 - 4AC}}{2A}$$

where the constants A, B and C (in general dependent upon k) have the following expressions.

$$A = \frac{1}{\nu_m^k}$$

$$B = 2\,\frac{c(1)}{\nu_m^k} - c(1) + \Theta$$

$$C = \frac{c^2(1)}{\nu_m^k} - \Theta\left(c(1) - \Theta\right) - \left(\frac{1}{\nu_{max}} + 1\right)\left(c(1) - \Theta\right)^2$$

with ν_m^k equal to the minimum value of ν relative to the (unique) region for which $\bar{c}^{(k)} = \Theta$ (see [4]).

Problem P 1

The value $n^{(k+1)}$ turns out to be the inverse of the value ν_n, which is the solution of the following min-max problem:

$$\min_{\hat{\nu} \le \nu_n \le \nu_R} \max\left(\eta_\psi(\nu_n), \eta_r(\nu_n)\right)$$

where

$$\eta_\psi(\nu_n) = 1 + \left(\frac{c(1)}{\Theta} - 1\right)\left(1 - \frac{\nu_n}{\nu_{max}}\right) \quad \text{for} \quad \hat{\nu} \le \nu_n \le \nu_R$$

$$\eta_r(\nu_n) = \begin{cases} \left(1 - \dfrac{\nu_{max} - \nu_n}{\nu_n}\left(\dfrac{c(1)}{c_m^{(k)}} - 1\right)\right)^{-1} & \text{for} \quad \hat{\nu} \le \nu_n < \nu_F \\[2ex] \dfrac{c_m^{(k)}}{\bar{\bar{c}}^{(k)}} & \text{for}^{(*)} \quad \nu_F \le \nu_n < \max\left\{\nu_F, \nu_P\right\} \\[2ex] \left(1 - \left(\dfrac{c(1)}{c_m^{(k)}} - 1\right)\left(\dfrac{\nu_n}{\nu_m^k(1+\nu_n)} - 1\right)\right)^{-1} & \text{for} \quad \max\left\{\nu_F, \nu_P\right\} \le \nu_n \le \nu_R \end{cases}$$

with $\hat{\nu}$, ν_F, ν_P and ν_R defined as follows.

Letting

$$\nu_Q = \nu_{max} \frac{c(1) - c_m^{(k)}}{c(1) - \Theta},$$

then

(*) max (a, b) = a if b is not defined.

$$\nu_F = \min\left\{ \frac{\nu_{max}\nu_m^k}{2}\left(1+\sqrt{1+\frac{4}{\nu_{max}\nu_m^k}}\right), \nu_{max}\left(1-\frac{c_m^{(k)}-\bar{\bar{c}}^{(k)}}{c(1)-\bar{\bar{c}}^{(k)}}\right)\right\}$$

$$\nu_R = \left[\frac{c(1)-c_m^{(k)}}{\nu_m^k\ c(1)-\Theta} \right]^{-1}$$

(where $[x]$ denotes the integer part of x)

$$\nu_P = \left(1-\left(1-\frac{\bar{\bar{c}}^{(k)}-c_m^{(k)}}{c(1)-c_m^{(k)}}\right)\right)^{-1} - 1$$

$$\hat{\nu} = \max\left\{ [[1/\nu_Q]]^{-1}, [[(1/\nu_F)+1]]^{-1} \right\}$$

(where $[[x]]$ denotes the maximum integer lower than x). ▨

Problem P 2

Assume k evaluations of $c(\cdot)$ have been made and let the N_k regions $R_i^{(k)}$ be ordered for increasing values of ν_n. Then, $N_k = i^{*(k)}$. Moreover let ν^i be the maximum value of ν_n belonging to $\mathcal{R}_i^{(k)}$, $i = 1,\dots,N_k-1$ and let $\mathcal{C}_{1/\nu^i}^{(k)} = \left[\bar{c}_i^{(k)}, \hat{c}_i^{(k)}\right]$.

The value $n^{(k+1)}$ turns out to be the inverse of ν, which is the solution of the following problem:

$$\min_{1\le i\le N_k} H(\nu^i)$$

where

$$H(\nu^i) = \frac{c_m^{(k)}}{\Theta} - \frac{1}{2\Theta}\,\frac{c_m^{(k)} - c_i^{(k)}}{\hat{c}_i^{(k)} - \bar{c}_i^{(k)}}\,, \qquad i = 1,\dots,N_k - 1$$

while $H(\nu^{N_k})$ and ν^{N_k} are the solutions of the following sub-problem (with ν_R defined as in P1):

$$H(\nu^{N_k}) = \min_{\nu_m^k < \nu_n \le \nu_R} K(\nu_n)$$

with $K(\nu_n)$ defined on the interval $\left(\nu_m^k, \nu_R\right]$ by

$$K(\nu_n) = \frac{1}{c(1) - \hat{c}_\alpha}\left\{\frac{1}{2\Theta}\left(2c_m^{(k)} c_\psi - \left(c_m^{(k)}\right)^2 - \hat{c}_\alpha^2\right) + \frac{c_m^{(k)}(c(1) - c_\varrho)}{c'_\alpha} + \right.$$

$$\left. + \frac{c_m^k \nu_n}{\nu_{max}}\, \lg \frac{\left(\frac{\nu_n}{\nu_{max}} - 1\right) c(1) + c_\varrho}{\left(\frac{\nu_n}{\nu_{max}} - 1\right) c(1) + c_\psi}\right\}.$$

Furthermore, c_ψ, c_ϱ, $\hat{c}_\alpha$ and c'_α are given by

$$c_\psi(\nu_n) = \Theta + (c(1) - Q)\left(1 - \frac{\nu_n}{\nu_{max}}\right) \qquad \text{for} \quad \nu_m^k < \nu_n \le \nu_R$$

$$c_\varrho(\nu_n) = \min\left\{\bar{\bar{c}}^{(k)} + \left(c(1) - \bar{\bar{c}}^{(k)}\right)\left(1 - \frac{\nu_n}{\nu_{max}}\right),\right.$$

$$\left., c(1) - \frac{c(1) - c_m^{(k)}}{1 + \nu_n} \cdot \frac{\nu_n^2}{\nu_{max}\,\nu_m^k}\right\} \qquad \text{for} \quad \nu_m^k < \nu_n \le \nu_R$$

$$\hat{c}_\alpha = \begin{cases} c_m^{(k)} - \left(c(1) - c_m^{(k)}\right)\left(\dfrac{\nu_n}{\nu_m^k} - 1\right) & \text{for} \quad \nu_m^k < \nu_n < \nu_R \\ \Theta & \text{for} \quad \nu_n = \nu_R \end{cases}$$

$$c'_\alpha = \begin{cases} \bar{\bar{c}}^{(k)} & \text{for} \quad \nu_m^k < \nu_n < \nu_z \\ c_m^{(k)} - \left(c(1) - c_m^{(k)}\right)\left(\dfrac{\nu_n}{\nu_m^k(\nu_n+1)} - 1\right) & \text{for} \quad \nu_z \leq \nu_n \leq \nu_R \end{cases}$$

with

$$\nu_z = \max\left\{\left[\left[1/\nu_m^k\right]\right]^{-1}, \ \nu_m^k \ \frac{c(1) - \bar{\bar{c}}^{(k)}}{c(1) - c_m^{(k)} - \nu_m^k\left(c(1) - \bar{\bar{c}}^{(k)}\right)}\right\}. \quad ▨$$

Problem P3

The value $n^{(k+1)}$ turns out to be the inverse of the value of ν which is the solution of the following min-max problem:

$$\min_{d^* \leq \nu_n \leq g} \max\left\{\bar{\eta}(\nu_n), \tilde{n}(\nu_n)\right\} \tag{5.2}$$

where

$$\bar{\eta}(\nu_n) = 1 + \frac{\left(c(1) - \bar{c}^{(k)}\right)(g - \nu_n)}{g \ \bar{c}^{(k)}} \quad \text{for } d^* \leq \nu_n \leq g$$

$$\tilde{\eta}(\nu_n) = \begin{cases} c_m^{(k)} \left(c_m^{(k)} - \left(c(1) - c_m^{(k)}\right) \dfrac{g - \nu_n}{\nu_n} \right)^{-1} & \text{for } \nu_n = d^* \\[2ex] \dfrac{c_m^{(k)}}{\bar{\bar{c}}^{(k)}} & \text{for}^{(*)}\ d^* < \nu_n < \max\{b, f\} \\[2ex] c_m^{(k)} \left(\bar{c}^{(k)} - \left(c(1) - \bar{c}^{(k)}\right) \left(\dfrac{\nu_n}{g(1-\nu_n)} - 1 \right) \right)^{-1} & \text{for}^{(*)}\ \max\{b, f\} \le \nu_n \le g \end{cases}$$

with g , f , b and d^* defined as follows:

g = maximum value of ν_n relative to the region where the lower-bound $\bar{c}^{(k)}$ is equal to max $\left\{ \nu_n : n \in \mathcal{N}^{(k)}_{i^*(k)} \right\}$

$$f = \frac{g^2\left(c(1) - c_m^{(k)}\right) + \sqrt{g^2\left(c(1) - c_m^{(k)}\right)^2 + 4g^2\left(c(1) - \bar{c}^{(k)}\right)\left(c(1) - c_m^{(k)}\right)}}{2\left(c(1) - \bar{c}^{(k)}\right)}$$

$$d^* = \left(1 + \left[\left[\max\left\{ f^{-1}, \frac{c(1) - \bar{\bar{c}}^{(k)}}{g\left(c(1) - c_m^{(k)}\right)} \right\} \right]\right] \right)^{-1}.$$

If eq. (5.2) yields more than one value of ν_n , then $\frac{1}{n^{(k+1)}}$ is the maximum of these values.

<u>Problem P 4</u>

Let
$$\bar{\nu} = \frac{-Y - \sqrt{Y^2 - 4XZ}}{2X}$$

(*) $\max(a, b) = a$ if b is not defined.

with

$$X = \frac{c(1)-\bar{c}^{(k)}}{\nu_{max}^{(k)}}\left(c(1)+\frac{c(1)-\bar{c}^{(k)}}{\nu_{max}^{(k)}}\right)$$

$$Y = -c(1)\left(c(1)-\bar{c}^{(k)}\right)\left(1+\frac{2}{\nu_{max}^{(k)}}\right)-\bar{c}^{(k)}\left(c(1)-c_m^{(k)}\right)$$

$$Z = c^2(1)-\bar{c}^{(k)}c_m^{(k)}$$

where $\nu_{max}^{(k)}$ is the maximum value of ν_n relative to the unique region $R_1^{(k)}$.

If $\frac{1}{\nu}$ is an integer, then $n^{(k+1)} = \frac{1}{\nu}$. Otherwise, let

$$\nu' = \left[\frac{1}{\nu}\right]^{-1}$$

$$\nu'' = \left[\frac{1}{\nu}+1\right]^{-1}.$$

Then $n^{(k+1)}$ turns out to be equal to $\frac{1}{\nu'}$ if

$$c_m^{(k)}\,\bar{c}^{(k)} < \left\{\bar{c}^{(k)} - \left(c(1)-\bar{c}^{(k)}\right)\left(\frac{\nu'}{\nu_{max}^{(k)}(1-\nu')}-1\right)\right\}$$
$$\left\{\bar{c}^{(k)} - \left(c(1)-\bar{c}^{(k)}\right)\frac{\nu''-\nu_{max}^{(k)}}{\nu_{max}^{(k)}}\right\}$$

while $n^{(k+1)} = \frac{1}{\nu''}$ if the opposite inequality holds. ▨

The problems P1 , P2 , P3 and P4 are concerned with the case where a lower bound $\Theta > 0$ of c° is a priori available. On the other hand, it often happens that $c^{\circ} \geq 0$ (i.e. $\Theta = 0$) (this is, for instance the case where $g(x,u) \geq 0$ in $X \times U$, i.e. the classical case of the positive semidefinite cost functions). When $\Theta = 0$, the here described procedure can not be used, since the index $\eta^{(k)}$ is not well defined. Nevertheless, this difficulty can be overcome by introducing an index, which is only formally different from $\eta^{(k)}$ and the solution of the secondary optimization problem turns out to be the one obtained via problems P1 and P2 for $\Theta \to 0$ (note that P3 and P4 do not depend upon Θ).

More in detail, for problem P1 the explicit solution is:

$$n^{(k+1)} = \left[\frac{c(1) - c_m^{(k)}}{\nu_m^k \, c(1)} \right],$$

while for problem P2 ν^{N_k} turns out to be equal to that value between $\left[\frac{c(1) - c_m^{(k)}}{\nu_m^k \, c(1)} \right]^{-1}$ and $\left(\left[\frac{c(1) - c_m^{(k)}}{\nu_m^k \, c(1)} \right] + 1 \right)^{-1}$,

for which $\lim_{\Theta \to 0} \Theta \, H(\nu_n)$ is minimum.

A similar procedure can be used when there are cycles with negative cost to time ratio (i.e. when a negative upper bound for c° is a priori known). The following secondary optimization problem should, then, be solved

(5.3) $$\max_{n \in \mathfrak{N}} \min_{c_n^{(k)} \in \mathscr{C}_n^{(k)}} \eta^{(k+1)}$$

instead of problem (3.3).

Finally, in the other cases the procedure can be the following. First evaluate $c(n_M)$ so that $\bar{c}^{(1)}$ is obtained. Then, if $\bar{c}^{(1)} \geq 0$, the problem is reduced to the one considered in the present section $(\Theta \geq 0)$. If $c(n_M) < 0$, criterion (5.3) must be applied. If, on the contrary, $c(n_M) > 0$ and $\bar{c}^{(1)} < 0$ (see Fig. 5), one should evaluate $c(\cdot)$ at the point $\nu_{n(2)}$ given by (see Fig. 5)

$$\nu_{n(2)} = \left[\left[\frac{1}{\nu'}\right]\right]^{-1} = \left[\left[\frac{c(1) - c(n_M)}{c(1)\nu_{min}}\right]\right]^{-1}.$$

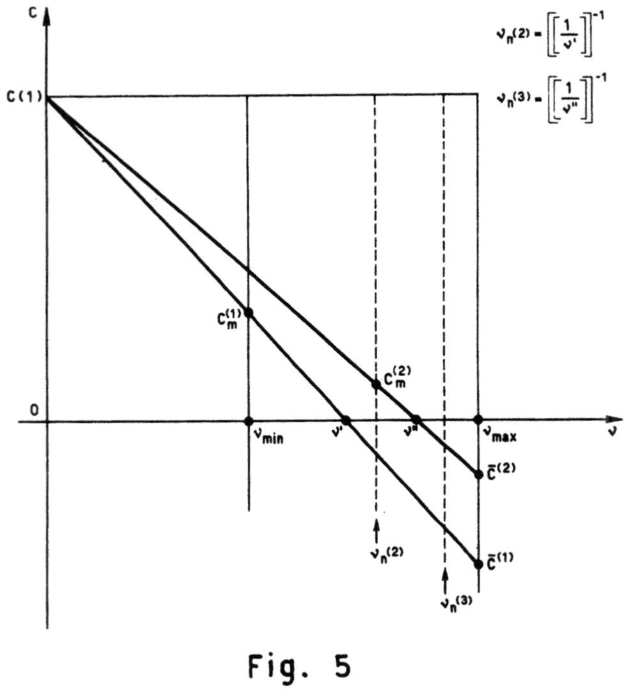

Fig. 5

Then, if $c_m^{(2)} > 0$ and $\bar{c}^{(2)} < 0$, a further evaluation of $c(\cdot)$ at $\nu_{n(3)}$ should be done, where (see Fig. 5) $\nu_{n(3)} = \left[\left[\frac{1}{\nu''}\right]\right]^{-1}$.

Obviously one should go on in the same way until $c_m^{(k)}$ and $\bar{c}^{(k)}$ turn out to be of the same sign (this does not happen only in the critical case $c^{\circ} = 0$).

REFERENCES

[1] K.S. Chang, " On the method of Computation for Optimal Periodic Processes" In this volume.

[2] A. Locatelli, S. Rinaldi, " An algorithm for the Determination of the Optimal Frequency of Periodic Processes" Automatica, Vol. 6, pp. 787-794, 1970

[3] C. Maffezzoni, S. Rinaldi, "Periodic Optimization of Discrete Systems", Ricerche di Automatica, Vol. 2, pp. 253-273, 1972.

[4] C. Maffezzoni, S. Rinaldi, "Periodic Optimization of Discrete Systems", Int. Rep. LCA 71-19, Istituto di Elettronica ed Elettrotecnica, Politecnico di Milano.

KUN S. CHANG (*)

NECESSARY AND SUFFICIENT CONDITIONS FOR OPTIMALITY

(*) Department of Chemical Engineering, University of Waterloo, Waterloo, Ontario, Canada

1. Introduction

For the last few years, an enormous number of research publications on periodic processes have appared in the literature (see Fjeld [1] , Guardabassi, Locatelli and Rinaldi [2] for the publications up to 1971). As the theory has reached a certain degree of growth and perfection, certain necessary and sufficient conditions have become available. Because of the peculiar behaviour of periodic processes, however, any extension of necessary and sufficient conditions of optimality for non-periodic processes to periodic processes is neither trivial nor straight-forward. The main difficulty is in the matching of the periodicity boundary conditions. This brings us to the problem of necessary and sufficient conditions for the existence and uniqueness of periodic solutions, a problem which was not severe in non-periodic cases. With this in mind, certain necessary and sufficient conditions for the existence and uniqueness of periodic solutions are first presented. Then the necessary conditions for optimality and the sufficient conditions, in particular the Hamilton-Jacobi theory, are covered. Certain conditions of relaxed controls are also briefly treated in the last section.

2. Conditions for Existence and Uniqueness of Periodic Solutions

Let the process be described by

$$\frac{dx}{dt} = f(x(t), u(t)) \tag{2.1}$$

with the boundary condition

$$x(0) - N(x(t_f)) = 0 \tag{2.2}$$

where $x \in E^n$, $u \in E^m$ and t_f are the state vector, the control vector, and a specified final time, respectively. For any arbitrary control $u(t)$, there is no guarantee that there exists a solution of (2.1) in the sense that the solution satisfies the condition (2.2). In addition, even if there exists a solution satisfying (2.2), the solution may not be unique. This is the source of some additional difficulties in the control of periodic processes. At the present time, there is no universal necessary and sufficient conditions which can be applied to general nonlinear systems of the type (2.1) and (2.2) to guarantee the uniqueness or at least the existence of a solution. However, there is a set of necessary and sufficient conditions for the system of equations where the state vector appears linearly or almost linearly, and a set of sufficient conditions for some nonlinear systems. These conditions were established mainly by Bounitzky, Massera, Levinson and Knobloch, and are well presented in Hartman [3]. Also a rigorous treatment of the theory of oscillations can be found in the work of Halanay [4]. From the computational point of view, Falb and de Jong's work [5] on successive approximation methods is useful. We begin first by treating linear equations. It is to be noted that whenever a function is referred to as " t_f -periodic", it means that the function is periodic with respect to time with period t_f.

2.1. Linear Equations

Let the control system be discribed by

$$\dot{x}(t) = A(v(t))\,x(t) + f(w(t)) \tag{2.3}$$

where A is an $n \times n$ matrix and f is an n-dimensional vector function, and v and w are p- and q-dimensional control vectors ($p + q = m$), respectively. We assume a general linear boundary condition of the type

$$x(0) - N\,x(t_f) = 0 \tag{2.4}$$

where N is an $n \times n$ nonsingular constant matrix which is, in many cases, the identity matrix I if the process state is t_f-periodic. When $v(t)$ and $w(t)$ are known as functions of t for $t \in [0, t_f]$, these functions can be substituted into (2.3) to obtain

(2.5) $$x = A(t)x + f(t).$$

Here we have slightly abused our notation and used the same A and f as in (2.3). It is obvious that if both $v(t)$ and $w(t)$ are t_f -periodic, i.e., $v(t+t_f) = v(t)$, $w(t+t_f) = w(t)$, then $A(t)$ and $f(t)$ are both t_f -periodic, i.e., $A(t+t_f) = A(t)$, $f(t+t_f) = f(t)$. We also write the corresponding homogeneous linear equation of (2.5)

(2.6) $$\dot{x} = A(t)x.$$

Theorem 2.1.

Let $A(t)$ be continuous for $t \in [0, t_f]$ and let $\Phi(t)$ be a fundamental matrix of (2.6). Then a necessary and sufficient cordition for the homogeneous system (2.6) to have a nontrivial solution satisfying (2.4) is that $\Phi(0) - N\Phi(t_f)$ be singular. The number of linearly independent solutions of (2.6) and (2.4) is the number of linearly independent vectors c satisfying

(2.7) $$[\Phi(0) - N\Phi(t_f)]c = 0.$$

Proof

Since the general solution of (2.6) is $x(t) = \Phi(t)c$, substitution of this into (2.4) gives $[\Phi(0) - N\Phi(t_f)]c = 0$. This equation has a nontrivial solution if and olny if $\Phi(0) - N\Phi(t_f)$

is singular. The second statement in the theorem follows because rank $[\Phi(0) - N\Phi(t_f)]$ + nullity $= n$. (Q.E.D.)

Theorem 2.2.

Let $A(t)$ be continuous for $t \in [0, t_f]$. The nonhomogeneous equation (2.5) has a unique solution $x(t)$ satisfying (2.4) for every continuous $f(t)$ if and only if the homogeneous equation (2.6) has no nontrivial solution satisfying (2.4).

Proof

The general solution of (2.5) can be written as

$$x(t) = \Phi(t)c + \Phi(t)\int_0^t \Phi^{-1}(s)\, f(s)\, ds. \tag{2.8}$$

This solution satisfies the boundary condition (2.4) if and only if

$$[\Phi(0) - N\Phi(t_f)]\, c = N\Phi(t_f)\int_0^{t_f} \Phi^{-1}(s)\, f(s)\, ds. \tag{2.9}$$

We assume that (2.6) has no nontrivial solution. Then by Theorem 2.1, $[\Phi(0) - N\Phi(t_f)]$ is nonsingular. This enables us to solve for c and substitute the value of c into (2.8) to get the unique solution

$$x(t) = \Phi(t)\Big\{[\Phi(0) - N\Phi(t_f)]^{-1} N\Phi(t_f)\int_0^{t_f} \Phi^{-1}(s)\, f(s)\, ds + \int_0^t \Phi^{-1}(s)\, f(s)\, ds\Big\}. \tag{2.10}$$

Conversely, we assume that (2.5) has a unique solution satisfying (2.4). This unique solution must be (2.10) and therefore $[\Phi(0) - N\Phi(t_f)]$ cannot be singular implying by Theorem 2.1 that (2.6) has no nontrivial solution. (Q.E.D.)

When the unique solution exists as in Theorem 2.2, this can be computed explicitly. Let $\Phi(t,\tau)$ be the fundamental matrix (transition matrix) of (2.6) such that $\Phi(t,\tau) = \Phi(t,0)\Phi^{-1}(\tau,0)$ and $\Phi(0,0) = I$. Then the initial value $x(0)$ that satisfies (2.5) and (2.4) is given by

$$x(0) = \left[I - N\Phi(t_f,0)\right]^{-1} N \int_0^{t_f} \Phi(t_f,0)\Phi^{-1}(s,0)\, f(s)\, ds. \tag{2.11}$$

The unique solution equivalent to (2.10) is then

$$\begin{aligned} x(t) = \Phi(t,0)\left[I - N\Phi(t_f,0)\right]^{-1} N \Phi(t_f,0) \int_0^{t_f} \Phi^{-1}(s,0) f(s)\, ds + \\ + \Phi(t,0) \int_0^{t} \Phi^{-1}(s,0) f(s)\, ds. \end{aligned} \tag{2.12}$$

This can be written as

$$x(t) = \int_0^{t_f} G(t,s)\, f(s) ds \tag{2.13}$$

where the Green's function $G(t,s)$ is given by

$$\begin{aligned} G(t,s) = \Phi(t,0)\left[I - N\Phi(t_f,0)\right]^{-1} N \Phi(t_f,0)\Phi^{-1}(s,0) + \\ + \Phi(t,0)\Phi^{-1}(s,0) H(t-s). \end{aligned} \tag{2.14}$$

Here the Heaviside's step function $H(t-s)$ is defined as

$$H(t-s) = \begin{cases} 1 & \text{if } 0 \le s \le t \le t_f \\ 0 & \text{if } 0 \le t < s \le t_f . \end{cases} \tag{2.15}$$

The expression (2.14) can be simplified. From the expansion of the obvious relationship $[I - N\Phi(t_f,0)]^{-1}[I - N\Phi(t_f,0)] = I$, one obtains

$$[I - N\Phi(t_f,0)]^{-1} N\Phi(t_f,0) + I = [I - N\Phi(t_f,0)]^{-1}. \tag{2.16}$$

Also from the expansion of the identity $[I - N\Phi(t_f,0)]^{-1}[I - N\Phi(t_f,0)] = [I - N\Phi(t_f,0)][I - N\Phi(t_f,0)]^{-1}$, one obtains

$$[I - N\Phi(t_f,0)]^{-1} N\Phi(t_f,0) = N\Phi(t_f,0)[I - N\Phi(t_f,0)]^{-1} \tag{2.17}$$

When (2.16) and (2.17) are substituted into (2.14) and the property $\Phi^{-1}(t,s) = \Phi(s,t)$ is used, we have either

$$G(t,s) = \begin{cases} \Phi(t,0)[I - N\Phi(t_f,0)]^{-1}\Phi(0,s), & 0 \le s \le t \le t_f \\ \Phi(t,0)[I - N\Phi(t_f,0)]^{-1} N\Phi(t_f,s), & 0 \le t < s \le t_f \end{cases} \tag{2.18a}$$

or

$$G(t,s) = \begin{cases} \Phi(t,0)[I - N\Phi(t_f,0)]^{-1}\Phi(0,s), & 0 \le s \le t \le t_f \\ \Phi(t,0) N\Phi(t_f,0)[I - N\Phi(t_f,0)]^{-1}\Phi(0,s), & 0 \le t < s \le t_f . \end{cases} \tag{2.18b}$$

Theorem 2.3.

If the solution is unique as in Theorem 2.2, then there exists a constant K , independent of $f(t)$, such that

$$\|x(t)\| \le K \int_0^{t_f} \|f(s)\| ds \quad \text{for} \quad t \in [0, t_f] \tag{2.19}$$

where $\| \quad \|$ is the Euclidean norm.

Proof

Obvious from (2.13) and (2.18). Take

$$\sup_{t,s \in [0,t_f]} \|G(t,s)\| = K.$$

(Q.E.D.)

It is to be noted that in Theorems 2.1, 2.2, and 2.3, nothing is said about the periodicity of $A(t)$, $f(t)$ and $x(t)$. At this point, it is useful to review the classical Floquet theory in order to have a good understanding of the properties of a fundamental matrix $\Phi(t)$ of (2.6).

Floquet Theory

Let $A(t)$ be a continuous $n \times n$ matrix for all t and be t_f -periodic. Then to every fundamental matrix $\Phi(t)$ of (2.6), there exists a t_f -periodic nonsingular $n \times n$ matrix $P(t) = P(t + t_f)$ and a constant $n \times n$ matrix W such that

$$\Phi(t) = P(t)e^{Wt}. \tag{2.20}$$

Since any matrix is similar to its Jordan canonical form, we can assume without loss of generality that W

is already in that form. It must also be recalled that

$$\Phi(t + t_f) = \Phi(t) e^{Wt_f} \tag{2.21}$$

where the matrix e^{Wt_f} is called a period transformation matrix (matrix of monodromy). The eigenvalues of e^{Wt_f} are the characteristic multipliers of the system and all period transformation matrices of the same system have exactly the same eigenvalues (invariant property of the system). The eigenvalues of W are called the characteristic exponents of the system.

In view of the Floquet theory, Theorem 2.1 implies that if $N = I$ and $A(t)$ is t_f -periodic, the homogeneous system (2.6) has a nontrivial t_f -periodic solution if and only if $\left[I - e^{Wt_f}\right]$ is singular or at least one eigenvalues of e^{Wt_f} is equal to one. In Theorem 2.2, if $A(t)$ is t_f -periodic, alternative forms of the Green's function in (2.18) can be obtained. To accomplish this, we apply the Floquet forms (2.20) and (2.21) to the fundamental matrix $\Phi(t,0)$ in (2.18) and obtain from (2.18) either

$$G(t,s) = \begin{cases} P(t)\, e^{Wt}\left[I - N e^{Wt_f}\right]^{-1} e^{-Ws} P^{-1}(s), & 0 \le s \le t \le t_f \\ P(t)\, e^{Wt}\left[I - N e^{Wt_f}\right]^{-1} N e^{W(t_f - s)} P^{-1}(s), & 0 \le t < s \le t_f \end{cases} \tag{2.22a}$$

or

$$G(t,s)=\begin{cases} P(t)\,e^{Wt}\left[I-N\,e^{Wt_f}\right]^{-1}e^{-Ws}\,P^{-1}(s)\,, & 0\le s\le t\le t_f \\ P(t)\,e^{Wt}\,N\,e^{Wt_f}\left[I-N\,e^{Wt_f}\right]^{-1}e^{-Ws}\,P^{-1}(s), & 0\le t<s\le t_f\,. \end{cases}$$

(2.22b)

If both $A(t)$ and $f(t)$ are t_f -periodic in (2.5), the following theorem applies.

Theorem 2.4.

Let $A(t)$ be continuous and t_f -periodic. Then for a fixed continuous t_f -periodic $f(t)$, the nonhomogeneous equation (2.5) has a t_f -periodic solution if and only if (2.5) has at least one bounded solution for $t\ge 0$.

Proof See [3].

2.2. Nonlinear Equations

We now consider the nonlinear control system of the form:

(2.23) $$\dot{x}(t)=A(v(t))\,x(t)+f(x(t),w(t))$$

where v and w are again p -and q -dimensional control vectors. Again let the boundary condition (2.4) be imposed. When the known controls $v(t)$ and $w(t)$ are substituted into (2.23), it can be written, again with some abuse of notation, as

(2.24) $$\dot{x}=A(t)x+f(t,x)\,.$$

We suppose that the homogeneous linear system

$$\dot{x} = A(t)\,x \tag{2.25}$$

does not admit any solutions other than the trivial one satisfying (2.4). Then it was seen in Theorem 2.1 that the unique solution is in the form of (2.13) if f is not a function of x. Therefore, if $G(t,s)$ is the Green's function of (2.25), we form the nonlinear integral equation

$$x(t) = \int_0^{t_f} G(t,x)\, f\left(s, x(s)\right) ds. \tag{2.26}$$

We can prove (see Halanay [4]) that (2.24) and (2.26) are equivalent, i.e., a solution of (2.26) is a solution of (2.24) and vice versa. In an operator notation, (2.26) can be written as

$$x(t) = \Gamma\left(x(t)\right). \tag{2.27}$$

The problem of proving the existence of a solution is reduced to that of proving the existence of the fixed points of operator Γ. By the fixed point principle, a fixed point exists in a complete metric space if the operator Γ is a contraction mapping and this fixed point is unique. This is essentially a problem of uniform convergence over $[0, t_f]$ in the Banach space of continuous functions $x(t)$ with the norm $|x|$ defined by

$$|x| = \sup_{t \in [0,t_f]} \|x(t)\|. \tag{2.28}$$

We now have the following theorem.

Theorem 2.5.

Let $A(t)$ be continuous ɪnd t_f -periodic such that (2.25) has no nontrivial t_f -periodic solution. Let K be as in (2.19) in Theorem 2.3, where $N = I$ in (2.4). Let $f(t,x)$ in (2.24) be continuous for all (t,x) and t_f -periodic in t for fixed x, and satisfy a Lipshitz condition of the form

$$\|f(t,x_1) - f(t,x_2)\| \leq L \|x_1 - x_2\| \tag{2.29}$$

for all t, x_1, x_2 with a Lipschitz constant L such that $KLt_f < 1$. Then (2.24) has a unique t_f -periodic solution. In fact, if

$$M = \sup_{t \in [0, t_f]} \|f(t,0)\|,$$

it is sufficient to require that $f(t,x)$ be defined only for $\|x\| \leq r$ where

$$\frac{K M t_f}{1 - K L t_f} \leq r . \tag{2.30}$$

The proof by the fixed point theorem in the Banach space is given in [3]. A proof on a similar theorem is also given in [4]. In Theorem 2.5, if $\|f(t,x)\|$ is small, we can omit the Lipschitz condition (2.29) at the expense of losing uniqueness.

Theorem 2.6.

Let $A(t)$ and K be as in Theorem 2.5 and $N = I$. Let $f(t,x)$ be continuous for all t and $\|x\| \leq r$ and t_f -periodic in t for fixed x, and satisfy

$$K\, t_f \, \| f(t,x) \| \leq r \tag{2.31}$$

for $t \in [0, t_f]$, $\|x\| \leq r$. Then (2.24) has at least one t_f-periodic solution.

<u>Proof</u> See [3].

In view of Theorems 2.5 and 2.6, for finite K, L , M , there exists a t_f -periodic solution of (2.24) if t_f is sufficiently small, and in Theorem 2.5, the solution is unique.

Now in (2.24) if we let

$$f(t,0) = g(t) \tag{2.32}$$

$$F(t,x) = f(t,x) - f(t,0) \tag{2.33}$$

an alternative form of (2.24) is given by

$$\dot{x} = A(t)x + g(t) + F(t,x). \tag{2.34}$$

The corresponding nonlinear integral equation similar to (2.26) becomes

$$x(t) = \int_0^{t_f} G(t,s)\, g(s)\, ds + \int_0^{t_f} G(t,s) F(s, x(s))\, ds \tag{2.35}$$

where the first integral represents the unique solution (let this be $\emptyset(t)$) of (2.34) with $F(t,x) = 0$. The form of (2.34) is very useful in the application of control theory of periodic processes. For example, the Riccati differential equation can be cast into this form. Using this form, Bittanti, Locatelli, and Maffezzoni [6] have established a set of sufficient conditions for the existence of the t_f-periodic solution of the Riccati equation arising in the accessory minimum problem. Let the $n \times n$ matrix Riccati equation be

$$\dot{S} = -SD - D^T S + SBS - C \tag{2.36}$$

where D, B and C are continuous $n \times n$ t_f-periodic matrices and D^T is the transpose of D. The following periodic boundary condition of (2.4) type $(N = I)$ is imposed.

$$S(0) - S(t_f) = 0 . \tag{2.37}$$

If the Riccati equation (2.36) with (2.37) is cast into the form of (2.34) and (2.4) with $N = I$, then $A(t)$ is a function of $D(t)$, $g(t)$ is a function of $C(t)$, $F(t, x)$ is a function of $B(t)$ and $S(t)$, and they are all t_f-periodic. The new x is an $n \times n$ dimensional vector consisting of all the elements of S. The dimensions of A, g and F are now all enlarged to give an enlarged $G(t,s)$. Let β be $\sup_{t \in [0, t_f]}$ [the maximum eingenvalue of $B^T(t)\, B(t)]^{1/2}$. A set of sufficient conditions for the existence of t_f-periodic solution of (2.36) is given in the following theorem.

<u>Theorem 2.7.</u>

If no characteristic multiplier of (2.25) is equal to one and

$$\beta \leq \frac{1}{4\sqrt{n}\; K\, t_f |\emptyset|} \tag{2.38}$$

where

$$K = \sup_{t,s \in [0,t_f]} \| G(t,s) \| ,$$

then there exists at least one t_f -periodic solution of (2.36) satisfying (2.37).

Proof See [6] .

2.3. Equations with Positive Solutions

Let the nonlinear control system again be represented by

$$\dot{x}(t) = f(x(t), u(t)) \tag{2.1}$$

subject to the boundary condition (2.4) with $N = I$. When a t_f -periodic control $u(t)$, $t \in [0, t_f]$, is substituted into (2.1), we have

$$\dot{x} = f(t, x(t)) \tag{2.39}$$

where f is t_f -periodic for any fixed x (an abuse of notation again). In many physical processes, the state variables are non-negative or positive quantities. A set of sufficient conditions for the existence of non-negative and positive t_f -periodic solutions of (2.39) has been established by Krasnoselskii

[7] . A concise summary of his conditions is given in the work of Kielkiewicz [8] and Fjeld [1] .

Theorem 2.8.

Let $f(t,x)$ in (2.39) be t_f -periodic in t for any fixed x . If the condition

$$f_i(t,x)\big|_{x_i=0} \geq 0 \qquad (x_j \geq 0,\ j \neq i) \tag{2.40}$$

is satisfied for all i $(i=1,2,\ldots,n)$ and the condition

$$[x^\alpha]^T f \leq 0 \tag{2.41}$$

is satisfied when the sum $x_1 + x_2 + \ldots + x_n$ $(x_j \geq 0,\ j=1,2,\ldots,n)$ is sufficiently large, where $[x^\alpha]^T = (x_1^\alpha\ x_2^\alpha\ \ldots\ x_n^\alpha)$ for an arbitrary fixed number $\alpha \in (-1,\infty)$; then the system (2.39) has at least one non-negative t_f -periodic solution $(x \geq 0)$ satisfying (2.4) with $N = I$.

Proof See Krasnoselskii [7] .

In Theorem 2.8, if additional conditions are imposed, (2.39) has a unique uniformly asymptotically stable positive $(x > 0)$ t_f -periodic solution (see [1] and [8]). These conditions are (i) $f(t,0) \not\equiv 0$ (ii) $f(t,x)$ has continuous partial derivatives with respect to all arguments and (iii)

$$f - f_x x \geq 0 \tag{2.42}$$

over $[0, t_f]$ for $0 < x_j$ $(j=1,2,\ldots,n)$ $< \infty$ where $(f_x)_{ij} = \dfrac{\partial f_i}{\partial x_j}$ for the ith row and the jth column. It is to be noted that the condition (2.41) comes from the existence of a Lyapunov-like function for the system (2.39).

2.4. Extended Floquet Theory

So far, in solving the homogeneous linear systems (2.6) and (2.25), we assumed that $A(t)$ is continuous. However, if the t_f-periodic control $v(t)$ is piecewise continuous over $[0, t_f]$ then $A(t)$ is only piecewise continuous over $[0, t_f]$. Then a fundamental matrix $\Phi(t)$ is discontinuous and the discontinuities in $\Phi(t)$ are transferred to the periodic matrix $P(t)$ in (2.20). The evaluation of $\Phi(t)$ for this case in the extended sense of Floquet theory has been given by Lee [9]. In particular, if $v(t)$ is piecewise constant over $[0, t_f]$, say, $v^1, v^2, \ldots, v^r$ over the time intervals $[0, t_1], [t_1, t_2], \ldots, [t_{r-1}, t_f]$, respectively, then $A(v^1), A(v^2), \ldots, A(v^r)$ are all piecewise constant over the same time intervals and e^{Wt_f} in (2.21) can be evaluated by

$$e^{Wt_f} = e^{A(v^r)T_r} e^{A(v^{r-1})T_{r-1}} \cdots e^{A(v^1)T_1} \tag{2.43}$$

where $T_1 = t_1 - 0$, $T_i = t_i - t_{i-1}$, $T_r = t_f - t_{r-1}$.
In addition, if $A(v^i)$ and $A(v^j)$ are commutative for all i and j, we have

$$e^{Wt_f} = \exp\left[\sum_{i=1}^{r} A(v^i)T_i\right]. \tag{2.44}$$

3. Necessary Conditions for Optimal Periodic Processes

An extensive list of publications dealing with continuous periodic systems has been compiled by Fjeld [1] and more recently by Guardabassi, Locatelli, and Rinaldi [2] . For periodic processes described by ordinary differential equations, Horn and Lin [10] have derived the necessary conditions for optimality and Bailey [11] has yielded the necessary conditions from a direct application of Pontryagin's principle (for a distributed-parameter system, see Fjeld and Kristiansen [12]).

We consider the process whose dynamics is given by

$$\frac{dx}{dt} = f(x(t), u(t)) \tag{3.1}$$

with the boundary condition

$$x(0) = N(x(t_f)) \tag{3.2}$$

where t_f is the final time and N is a continuously differentiable function with respect to its argument and the Jacobian $\neq 0$. The state x is n-dimensional and the control u is m-dimensional, and $u \in \Omega$ where Ω is the set of piecewise continuous functions. The output vector may be given. However, we assume that the state is identical to the output. The performance index to be minimized is given by

$$J = \frac{1}{t_f}\int_0^{t_f} L(x,u)\,dt . \tag{3.3}$$

We now introduce the following transformation

$$\tau = \frac{t}{t_f} \quad \text{or} \quad dt = t_f\, d\tau . \tag{3.4}$$

The substitution of (3.4) into (3.1) - (3.3) yields

$$\frac{dx}{d\tau} = F(x,u,t_f) \tag{3.5}$$

where

$$F(x,u,t_f) \triangleq t_f\, f(x,u) \tag{3.6}$$

$$x(0) = N(x(1)) \tag{3.7}$$

and

$$J = \int_0^1 L(x,u)\,d\tau . \tag{3.8}$$

We use (3.5) - (3.8) and try to find $u(\tau) \in \Omega$, $\tau \in [0,1]$ such that J is minimum. Define a Hamiltonian function by

$$H(x,u,\lambda,t_f) \triangleq L(x,u) + \lambda^T F(x,u,t_f) \tag{3.9}$$

where T stands for transpose.

The state and the boundary constraints are adjoined to the performance index by Lagrange multipliers $\lambda(t)$ and ν , respectively,

$$\bar{J} = \nu^T\left[N(x(1)) - x(0)\right] + \int_0^1 \left[H - \lambda^T \dot{x}\right] d\tau \tag{3.10}$$

where $\dot{x} = \dfrac{dx}{d\tau}$.

If the constraints are exactly met, obviously $\bar{J} = J$. The first variation of $\bar{J}$ is derived for free t_f as

$$\delta\bar{J} = \left[\lambda^T(0) - \nu^T\right]\delta x(0) - \left[\lambda^T(1) - \nu^T \frac{\partial N}{\partial x(1)}\right]\delta x(1) +$$

$$+ \delta\nu^T\left[N(x(1)) - x(0)\right] +$$

$$+ \int_0^1 \Big[H_x^T\,\delta x + H_u^T\,\delta u + H_\lambda^T\,\delta\lambda + H_{t_f} dt_f +$$

$$+ \dot{\lambda}^T \delta x - \delta\lambda^T \dot{x}\Big] d\tau + o(\delta) \tag{3.11}$$

where $H_x^T \triangleq \left(\dfrac{\partial H}{\partial x_1}, \dfrac{\partial H}{\partial x_2}, \dots, \dfrac{\partial H}{\partial x_n}\right)$, T for transpose and $\lim\limits_{\delta \to 0} \dfrac{o(\delta)}{\delta} = 0$.

By setting $\delta\bar{J} = 0$ for arbitrary small variations in x, u, λ, t_f and ν, the following set of necessary conditions for optimality results:

$$\dot{x} = H_\lambda, \quad x(0) = N(x(1)) \tag{3.12}$$

$$\dot{\lambda} = -H_x, \quad \lambda(1) = \left(\frac{\partial N}{\partial x(1)}\right)^T \lambda(0) \tag{3.13}$$

$$H_u = 0 \tag{3.14}$$

$$\int_0^1 H_{t_f} d\tau = 0 . \tag{3.15}$$

The condition (3.15) is the stopping condition for t_f and can be rewritten for the present H in (3.9) as

$$\int_0^1 H \, d\tau = J . \tag{3.16}$$

If the optimal control cannot be obtained by the minimization of H by partial differentiation with respect to u (a constrained case) as in (3.14), then we must start again from (3.10). To do this, let the difference $\bar{J}(x + \Delta x, u + \Delta u, \lambda + \Delta\lambda, t_f + \Delta t_f) - \bar{J}(x, u, \lambda, t_f)$ be $\Delta\bar{J}$.

Then it can be shown by Rozonoér's method [13] that

$$\Delta\bar{J} = \int_0^1 \left[H(x, u + \Delta u, \lambda, t_f) - H(x, u, \lambda, t_f) \right] d\tau + \eta \tag{3.17}$$

where $|\eta|$ can be made arbitrarily small by choosing a sufficiently small time interval. Then $\Delta\bar{J} \geq 0$ implies $[H(x, u + \Delta u, \lambda, t_f) - H(x, u, \lambda, t)] \geq 0$. Therefore, H must be minimum for the optimal control. Thus the condition (3.14) can be replaced by that of H being minimum. This is the minimum principle for the system.

Differentiation of H with respect to τ gives

$$\dot{H} = H_\tau + H_x^T \dot{x} + H_u^T \dot{u} + H_\lambda^T \dot{\lambda} + H_{t_f} \dot{t}_f =$$

$$= H_\tau + H_u^T \dot{u} + (H_x^T + \dot{\lambda}^T) F = H_\tau + H_u^T \dot{u} . \tag{3.18}$$

Since H does not contain τ explicitly, $H_\tau = 0$.
If the minimum H is obtained by (3.14), then $H_u = 0$. On the other hand, if the minimum H is achieved by a control of bang-bang type, then $\dot{u} = 0$ almost everywhere. In either case, $\dot{H} = 0$ and thus

$$H(\tau) = \text{constant}, \ \tau \in [0,1]. \tag{3.19}$$

If t_f is free, from (3.16) and (3.19), we have

$$H = J . \tag{3.20}$$

The results are summarized in the following theorem.

Theorem 3.1.

Let $\hat{u}(\tau)$ be an admissible control, $\hat{u} \in \Omega$. Let $\hat{x}(\tau)$ be the solution of (3.5) and (3.7) corresponding to $\hat{u}$. In order that $\hat{u}$ be optimal, it is necessary that there exist a function $\hat{\lambda}$ such that for fixed t_f
(i) $(\hat{x}, \hat{\lambda})$ is the solution of the canonical system

$$\begin{aligned} \dot{\hat{x}} &= H_\lambda(\hat{x}, \hat{u}, \hat{\lambda}, t_f) \\ \dot{\hat{\lambda}} &= -H_x(\hat{x}, \hat{u}, \hat{\lambda}, t_f) \end{aligned} \tag{3.21}$$

satisfying the boundary conditions

$$\hat{x}(0) = N(\hat{x}(1))$$
$$\hat{\lambda}(1) = \left(\frac{\partial N}{\partial \hat{x}(1)}\right)^T \hat{\lambda}(0). \tag{3.22}$$

(ii) The function $H(\hat{x}, u, \hat{\lambda}, t_f)$ has an absolute minimum as a function of u over Ω at $u = \hat{u}$ almost everywhere over $[0,1]$, i.e.,

$$\min_{u \in \Omega} H(\hat{x}, u, \hat{\lambda}, t_f) = H(\hat{x}, \hat{u}, \hat{\lambda}, t_f) \tag{3.23}$$

for $\tau \in [0,1]$.

(iii) $H(\hat{x}, \hat{u}, \hat{\lambda}, t_f)$ = constant for $\tau \in [0,1]$.

If t_f is free, the conditions (i) and (ii) are again satisfied with the optimal final time $\hat{t}_f$ in place of t_f where $\hat{t}_f$ is chosen such that the condition

$$H(\hat{x}, \hat{u}, \hat{\lambda}, \hat{t}_f) = J(\hat{x}, \hat{u})$$

is satisfied for $\tau \in [0,1]$.

Suppose that there is an additional k -dimensional integral constraint equation of the form

$$\int_0^1 h(x(\tau), u(\tau))\, d\tau = 0. \tag{3.24}$$

For this case, we define an additional state z by

$$\frac{dz}{d\tau} = h(x(\tau), u(\tau)) \tag{3.25}$$

$$z(0) = z(1) \tag{3.26}$$

and enlarge the dimension of x in (3.5) from n to $(n+k)$ by appending (3.25), (3.26) to (3.5), (3.7) and by redefining x. Then the same procedure can be followed. In Theorem 3.1, if $N = I$, the principle is the minimum principle for the t_f-periodic process $(t_f = 1)$.

4. Sufficient Conditions for Optimal Periodic Processes

For periodic processes, a set of sufficient conditions for optimality based on a second variation method has been established by Bittanti, Locatelli, and Maffezzoni [6] . A set of sufficient conditions for free t_f has also been established in Chang [14] . For these conditions, refer to [14]. Sufficient conditions for the possible improvement of optimal steady state have been derived in Horn and Lin [10] , Bailey and Horn [15] , and Guardabassi [16] . Instead of repeating the conditions here, we shall review a different set of sufficient conditions for optimality. These conditions are based on the Hamilton-Jacobi theory (for details, see Maffezzoni's work [17]). The essential part of the result will be stated here for the system

$$\frac{dx}{d\tau} = F(x,u,t_f) \tag{4.1}$$

with

$$x(0) = x(1). \tag{4.2}$$

Let $\Phi(\tau; 0, x(0), u(\tau))$ be the transition function of (4.1) from $x(0)$ to $x(\tau)$. Let the Hamiltonian function H be defined by (3.9) and let the set R be defined by $R \triangleq [0,1] \times X$ where X is the state set. If H is regular relative to R, define the Hamilton-Jacobi equation of the control problem relative to R by

$$V_\tau(\tau,x) + H(x, V_x(\tau,x), u^*(\tau,x,V_x(\tau,x)),\tau) = 0 \tag{4.3}$$

where $u^*(\tau,x,V_x)$ is the H-minimal control relative to R.

<u>Theorem 4.1.</u>

Suppose that H is regular relative to R and that the transition function Φ is continuously differentiable on R. Let $\hat{u} \in \Omega$ such that

(i) $\hat{u}$ generates a t_f-periodic solution $\hat{x}$ in R (here $t_f = 1$).

(ii) There is a continuously differentiable solution $V(\tau,x)$ of the Hamilton-Jacobi equation (4.3) on R satisfying

$$V(0,x) - V(1,x) = c(1) \tag{4.4}$$

for all $x \in X$ where $c(\tau)$ is a scalar function,

and

(4.5) $$\hat{u}(\tau) = u^*(\tau, \hat{x}(\tau), V_x(\tau, \hat{x}(\tau))$$

where $\hat{x}(\tau) = \Phi(\tau; 0, x(0), \hat{u}(\tau))$ and $\hat{x}(0) = \hat{x}(1)$.

Then $\hat{u}$ is optimal relative to R.

Proof See [17]. It turns out that the constant $c(1)$ represents the minimum J relative to R.

5. Relaxed Control of Periodic Processes

5.1. Relaxed Controls

The theory of " relaxed controls" is based on the concept of " generalized curves" which was first proposed by Young [18] and further developed by McShane [19-22] in the study of the classical calculus of variations. Warga [23,24] treated the subject as relaxed variational problems and established the necessary conditions for minimality of the problems. For certain non-convex problems, an optimal control may not exist in the conventional sense but may exist in the relaxed control sense. Noting this, the word " sliding regimes" was used by Filippov [25]. In the relaxed control, the permissible set of choices for the derivative of the trajectory is enlarged from a non-convex set to the convex closure of that set. Incessant switchings of infinite rapidity occur in the control and, therefore, it was called " chattering controls" by Lee and Markus [26] . McShane [27] and Warga [28] have further enriched the development. A concise summary of the theory is given in Horn and Bailey [29] and also in Fjeld [1] .

Let the system be described by

$$\frac{dx}{dt} = f(x(t),u(t)) \tag{5.1}$$

with the initial and terminal conditions $x(0)$ and $x(t_f)$ where the n-dimensional state $x \in X$ and the m-dimensional control $u \in U$. U may be a finite set. Let f be non-convex and be defined for all $x \in X$, $u \in U$, and $t \in [0,\infty)$. We assume that an appropriate form of the performance index is given. An equivalent form of (5.1) is

$$x(t) = x(0) + \int_0^t f(x(s),u(s))\,ds . \tag{5.2}$$

Let M be a mean value operator with the properties: $M[c_1\Phi_1 + \Phi_2] = c_1 M[\Phi_1] + M[\Phi_2]$, $M[1] = 1$, and $M[\Phi] \geq 0$ if $\Phi \geq 0$ and Φ continuous over U . Now we replace the intergrand of (5.2) by $M[f]$ to get

$$x(t) = x(0) + \int_0^1 M[f(x(s),u(s))]\,ds . \tag{5.3}$$

In particular, M can be a convex closure operator. Thus, the problem with (5.2) is the original problem and one with (5.3) is the relaxed problem. An important point is that according to Warga [23] , any relaxed admissible curve of (5.3) can be uniformly approximated by the curves of (5.2). Now a theorem on convexity applies. According to the theorem of Caratheodory, in an m-dimensional vector space, each point in the convex hull of a set can be obtained by a convex combination of at most $m+1$ points of the original set, and m points if the original set is connected. Let the points $u^1, u^2, \ldots, u^k$ of U be the chattering basis of U, and during a short time interval $[t, t+\delta]$ the controls are set on u^1 for time $\alpha_1\delta, \ldots,$ on u^k for time $\alpha_k\delta$, $\alpha_j \geq 0$ $(j=1,2,\ldots,k)$, $\sum_{j=1}^{k}\alpha_j = 1$, then the integral of (5.3) from t to $t+\delta$ is nearly

$$\int_{t}^{t+\delta} \sum_{j=1}^{k} \alpha_j f(x(s), u^j)\, ds . \tag{5.4}$$

As δ becomes smaller, the control switchings become faster; then the control is called a " relaxed control".

5.2. Relaxed Steady States

The idea of relaxed control has been applied to periodic processes by Horn, Bailey, and Lin [15, 29–35]. The condition for a relaxed steady state in a chemical process was also found by Fjeld [36]. A " relaxed state" $x_r(t)$ is defined to be the solution of (5.3) generated by a relaxed control u_r. An asymptotically stable " relaxed steady state" x_{rs} is said to exist if $\lim_{t\to\infty} x_r(t) = x_{rs}$ and x_{rs} is constant. Let the set of all steady state solutions of (5.1) for all $u \in U$ be denoted by $x_s[U]$. If $x_{rs} \notin x_s[U]$, then x_{rs} is called a "proper relaxed steady state". A necessary and sufficient condition for the existence of a relaxed steady state is that there exists at least one $x_{rs} \in X$ such that

$$\sum_{j=1}^{k} \alpha_j f(x_{rs}, u^j) = 0 \tag{5.5}$$

where $\sum_{j=1}^{k} \alpha_j = 1$ and k need never be larger than $n+1$ (n is the dimension of x).

5.3. Systems without Proper Relaxed Steady State

Not all systems have a proper relaxed steady state. For example, the following system does not have a proper relaxed steady state (see Fjeld [1]). This was also observed by Bankoff and Chang [37] for a distributed-parameter system.

$$\dot{x} = f(x(t)) + u(t) \tag{5.6}$$

$$J = \frac{1}{t_f} \int_0^{t_f} g(x(t))\, dt \tag{5.7}$$

where the function f , and g are differentiable, and the set of admissible controls, U , is compact and convex.

REFERENCES

[1] M. Fjeld, " Stability and Control of Periodic Processes", Rep. 71-52-W, Division of Automatic Control, The Technical University of Norway, Trondheim (April 1971).

[2] G. Guardabassi, A. Locatelli, and S. Rinaldi, " The Status of Periodic Optimization of Dynamical Systems", Int. Rep. LCA 71-20, Istituto di Elettrotecnica ed Elettronica, Politecnico di Milano (Nov. 1971).

[3] P. Hartman, "Ordinary Differential Equations", John Wiley and Sons, Inc. (1964).

[4] A. Halanay, " Differential Equations", Academic Press (1966).

[5] P.L. Falb and J.L. de Jong, " Some Successive Approximation Methods in Control and Oscillation Theory", Academic Press, (1969).

[6] S. Bittanti, A. Locatelli and C. Maffezzoni, " Second Variation Methods in Periodic Optimization", Int. Rep. LCA 71-17, Istituto di Elettrotecnica ed Elettronica, Politecnico di Milano (Oct. 1971).

[7] M.A. Krasnoselskii, " Positive Solutions of Operator Equations" P. Noordhoff Ltd. (1964).

[8] M. Kielkiewicz, " Existence of Positive Periodic Solutions of Certain Ordinary Differential Equations", Electronics Letters, 6, N°15, 481-482 (1970).

[9] I. Lee, " On the Theory of Linear Dynamic System with

Periodic Parameters", Inf. Control, 6, 265-275 (1963).

[10] F.J.M. Horn and R.C. Lin, " Periodic Processes: Variational Approach", Ind. Eng. Chem. Proc. Des. Dev., 6, 21-30 (1967).

[11] J.E. Bailey, " Necessary Conditions for Optimality in a General Class of Nonlinear Mixed Boundary Value Control Problems", (to appear in Int. J. Control).

[12] M. Fjeld and T. Kristiansen, " Optimization of a Nonlinear Distributed Parameter System using Periodic Boundary Control", Int. J. Control, 10, 601-624 (1969).

[13] L.I. Rozonoér, " L.S. Pontryagin Maximum Principle in the Theory of Optimum Systems, I", Autom. Rem. Control, 20, 1288-1302 (1960).

[14] K.S. Chang, " Second Variation for Periodic Optimization Problems", In this volume.

[15] J.E. Bailey and F.J.M. Horn, " Comparison between Two Sufficient Conditions for Improvement of an Optimal Steady-State Process by Periodic Operation", J. Opt. Theory Appl., 7, 378-384 (1971).

[16] G. Guardabassi, " Steady-State versus Periodic Optimization: A Circle Criterion" , Int. Rep. LCA 71-13, Istituto di Elettrotecnica ed Elettronica, Politecnico di Milano (June 1971).

[17] C. Maffezzoni, " Hamilton-Jacobi Theory for a Periodic Control Problem" , Int. Rep. LCA 71-16, Istituto di Elettrotecnica ed Elettronica, Politecnico di Milano (Sept. 1971).

[18] L.C. Young, " Generalized Curves and the Existence of an

Attained Absolute Minimum in the Calculus of Variations", Compt. Rend. Soc. Sci. et Lettres Varsovie, Cl. III, 30, 212-234 (1937).

[19] E.J. McShane, " Generalized Curves", Duke Math. J., 6, 513 (1940).

[20] E.J. McShane, " Necessary Conditions for Generalized-Curve Problems of the Calculus of Variations", Duke Math. J., 7, 1 (1940).

[21] E.J. McShane, " Existence Theorem for Bolza Problems in the Calculus of Variations", Duke Math. J., 7, 28 (1940).

[22] E.J. McShane, " Some Existence Theorems in the Calculus of Variations. III Existence Theorems for Non-Regular Problems", Trans. Amer. Math. Soc., 45, 151 (1939).

[23] J. Warga, " Relaxed Variational Problems" , J. Math. Anal. Appl., 4, 111-128 (1962).

[24] J. Warga, " Necessary Conditions for Minimum in Relaxed Variational Problems", J. Math. Anal. Appl., 4, 129-145 (1962).

[25] A.F. Filippov, " On Certain Questions in the Theory of Optimal Control", SIAM J. Control, 1, 76-84 (1962).

[26] E.B. Lee and L. Markus, " Foundations of Optimal Control Theory", John Wiley and Sons, Inc. (1967).

[27] E.J. McShane, " Relaxed Controls and Variational Problems", SIAM J. Control, 5, 438-485 (1967).

[28] J. Warga, " Functions of Relaxed Controls", SIAM J., 5, 628-641 (1967).

[29] F.J.M. Horn and J.E. Bailey, " An Application of the Theorem of Relaxed Control to the Problem

of Increasing Catalyst Selectivity" , J. Opt. Theory Appl., 2, 441-449 (1968).

[30] F.J.M. Horn and J.E. Bailey, " Catalyst Selectivity under Steady-State and Dynamic Operation" , Ber. Bunsenges. Phys. Chemie, 73, 274-279 (1969).

[31] F.J.M. Horn and J.E. Bailey, " Catalyst Selectivity under Steady-State and Dynamic Operation: An Investigation of Several Kinetic Mechanisms", Ber, Bunsenges. Phys. Chemie, 74, 611 (1970).

[32] J.E. Bailey and F.J.M. Horn, " Improvement of the Performance of a Fixed Bed Catalytic Reactor by Relaxed Steady-state Operation", AIChE J., 17, 550-553 (1971).

[33] J.E. Bailey and F.J.M. Horn, " Cyclic Operation of Reaction Systems: The Influence of Diffusion on Catalyst Selectivity", Chem. Eng. Sci., 27, 109-119 (1972).

[34] J.E. Bailey, F.J.M. Horn, and R.C. Lin, " Cyclic Operation of Reaction Systems: Effect of Heat and Mass Transfer Resistance", AIChE J., 17, 818-825 (1971).

[35] J.E. Bailey, " Tangent Cone Analysis of Periodic Preceses" , AIChE J., 18, 459-461 (1972).

[36] M. Fjeld, " On Periodic Processes. A Preliminary Study and Propositions for Further Investigations" Rep. 67-54-D, Division of Automatic Control, The Technical University of Norway, Trondheim (April 1967).

[37] S.G. Bankoff and K.S. Chang, " Oscillatory Operation of Jacketed Tubular Reactors" , Ind. Eng. Chem. Fundamentals, 9, 301 (1970).

CONTENTS

Page

Preface... 3

G. Guardabassi, A. Locatelli, S. Rinaldi
"What is Periodic Optimization"........................ 5

P. Dorato
"Asymptotic Optimization in Finite Directed Cycles".. 17

E.L. Lawler
"Optimal Cycles in Graphs and the Minimal Cost-to-Time Ratio".. 37

A. Locatelli, P. Migliarese, S. Rinaldi
"Constrained Periodic Optimization of Finite Automata".. 61

F. Maffioli
"Optimal Hamiltonian Cycles: A Survey of Recent Results"... 87

S. Bittanti, G. Fronza, G. Guardabassi
"Optimality Conditions for Discrete Periodic Optimization Problems"..................................... 113

S. Bittanti, G. Fronza, G. Guardabassi
"Optimal Steady - State Vs Periodic Operation in Discrete Systems"...................................... 133

A. Locatelli, C. Maffezzoni, S. Rinaldi
"A Searching Procedure of the Optimal Frequency in Periodic Optimization"................................ 159

K. S. Chang
"Necessary and Sufficient Conditions for Optimality".. 183

MIX
Papier aus verantwortungsvollen Quellen
Paper from responsible sources
FSC® C105338

If you have any concerns about our products,
you can contact us on
ProductSafety@springernature.com

In case Publisher is established outside the EU,
the EU authorized representative is:
Springer Nature Customer Service Center GmbH
Europaplatz 3, 69115 Heidelberg, Germany

Printed by Libri Plureos GmbH
in Hamburg, Germany